Agent-based Models and Causal Inference

Agent-based Models and Causal Inference

Gianluca Manzo

Department of Sociology
Sorbonne University (SU)
Paris, France

Registered Office(s)
John Wiley & Sons, Inc., 111 River Street, Hoboken, NJ 07030, USA
John Wiley & Sons Ltd, The Atrium, Southern Gate, Chichester, West Sussex, PO19 8SQ, UK

Editorial Office
9600 Garsington Road, Oxford, OX4 2DQ, UK

For details of our global editorial offices, customer services, and more information about Wiley products visit us at www.wiley.com.

Wiley also publishes its books in a variety of electronic formats and by print-on-demand. Some content that appears in standard print versions of this book may not be available in other formats.

Library of Congress Cataloging-in-Publication Data
Names: Manzo, Gianluca, author.
Title: Agent-based models and causal inference / Gianluca Manzo.
Description: Hoboken, NJ : John Wiley & Sons, Inc., 2022. | Includes
  bibliographical references and index.
Identifiers: LCCN 2021032403 (print) | LCCN 2021032404 (ebook) | ISBN
  9781119704478 (hardback) | ISBN 9781119704454 (pdf) | ISBN 9781119704461
  (epub) | ISBN 9781119704492 (ebook)
Subjects: LCSH: Qualitative research--Methodology. | Social
  sciences--Methodology. | Multiagent systems. | Causation. | Inference. |
  Probabilities. | Multivariate analysis.
Classification: LCC H62 .M23596 2022 (print) | LCC H62 (ebook) | DDC
  300.72/1--dc23
LC record available at https://lccn.loc.gov/2021032403
LC ebook record available at https://lccn.loc.gov/2021032404

Cover image: Wiley
Cover design by Wiley

Set in 9.5/12.5pt STIXTwoText by Integra Software Services, Pondicherry, India
Printed and bound by CPI Group (UK) Ltd, Croydon, CR0 4YY

C9781119704478_240122

"Of Course, many of my colleagues will be found to disagree. For them, fitting models to data, computing standard errors, and performing significance tests is "informative", even though the basic statistical assumptions (linearity, independence of errors, etc) cannot be validated. This position seems indefensible, nor are the consequences trivial. Perhaps it is time to reconsider".

D. Freedman (1995: 19)

Contents

List of Acronyms

ABM: an Agent-based Model or Agent-based Modeling
ABMs: Agent-based Models
DAG(s): Directed Acyclic Graph(s)
IV(s): Instrumental Variable(s)
KISS (principle): Keep It Simple, Stupid
KIDS (principle): Keep It Descriptive, Stupid
RCTs: Randomized Control Trials (experiments)
SUTVA: Stable Unit Treatment Value Assumption

List of Table

Preface

Towards the end of 2020, I was doing a bibliography search for completing another book project and I came across a short piece by Peter Bearman titled "Notes for heuristics of discovery", published in 2018 in the *International Journal for Sociological Debate* (*Sociologica*). In the opening section of this reflexive paper, Berman explained that the most difficult task for him about a given piece of work is "knowing when it is finished". Then the readers learn that in fact, for most of his research, Bearman "waited years between finishing papers and submitting them to journals, essentially unchanged after years spent in a box, or file cabinet". Bearman explained that he finally realized that this recurrently happens to him because, in order to regard a project as finished, he needs to understand what its contribution could be. And finding the answer takes time, faces hesitation, and generates potential frustration.

When I discovered Bearman's piece, I felt at the same time excited and reassured: excited because Bearman's words precisely described my procrastination with closing the present book, reassured because Bearman's words helped me to understand the reason for this. I prepared indeed a first version of the manuscript between 2018 and 2019, submitted it for publication to Wiley in the fall of 2019, and only completed the final version almost one and a half years later. But, as I explain below, the history of the book spanned over a much larger temporal window. Thanks to Bearman's reflection, I understand now that the reason for this tortuous path was precisely that described by Bearman. I hesitated about the contribution that I wanted to make as well as about how to make it.

Causal inference is a hugely complex topic that has been treated in a highly technical manner by many statisticians, econometricians, and, more recently, computer scientists. Philosophers have also written extensively on causality in a very specialized way for a long time. As a reader of (some of) this literature, and as a sociologist with a strong interest in a specific computational method, i.e. agent-based modeling, I had the intuition that this method could contribute to causal reasoning, but, at the same time, I had the impression that this was not even an option worthy of discussion for many quantitative social scientists. It took time for me to understand that what I really wanted to write was a book that contributed to changing the perception of a given tool, i.e. agent-based computational modeling, which is not usually thought of as a tool for causal inference.

And it also took time for me to understand how, as a sociologist, I could have best reached this goal. A book on the philosophy of causality did not seem a good solution, if not because I am not a philosopher; moreover, an entirely abstract treatment of the issue would have probably been ineffective for empirically minded sociologists interested in causal inference. But I did not want either to write a technical book on this or that method for causal inference, nor on agent-based modeling itself. Plenty of competent treatments along these lines already exist, and I believe that this is a job for professional statisticians and computer scientists. Only gradually I came to realize that one alternative option was to connect agent-based computational modeling to theories of causality and mechanisms among philosophers of science and, at the same time, to compare this tool with more established experimental and observational methods for causal inference in statistics, economics, and computer science. A meta-theoretical but methodologically informed study at the intersection of different research fields and disciplines finally appeared to me a defensible intermediate stance between the philosophy of causality and pure techniques of causal inference. Thus the contribution that I finally thought I could have made with this book amounts to a new conceptual grid bridging scholarship on causality across several disciplines and helping us to understand why causal inference should be pursued at the intersection of a variety of formal methods.

But I needed several years, various academic events, and personal encounters to see my project in this way. Retrospectively, I can now say that causality actually was a longstanding interest of mine. When drafting the theoretical chapter of my PhD dissertation, back in December 2004, I came across Rom Harré's distinction between "successionist" and "generative" causality. The focus of that chapter changed a little during the following year; thus the discussion of this conceptual distinction disappeared from the version of the PhD dissertation that I finally submitted to the committee early in 2006. However, I went back to this couple of concepts in 2010 when I was preparing a contribution to the book on social mechanisms that Pierre Demeulenaere was editing at Cambridge University Press. Demeulenaere wanted to focus the chapters on causal explanation, and to its link with the concept of mechanism. To meet the editor's requirement, I tried to connect the paper that I had planned to submit for this book, which was about an agent-based model of relative deprivation, with the concept of "generative causality" that I found in Harré's writings some years earlier. This was a branching point: I realized that there was room to elaborate more systematically the role of agent-based computational modeling for causal inference.

The opportunity to deepen the analysis of this connection arrived a couple of years later. In particular, Isabelle Drouet invited me to give one of the keynotes at a conference on "Causality and experimentation in the sciences" that she was organizing at Sorbonne University. The conference took place in July 2013, and this was a major step in my thinking on causality. On the one hand, the talks I listened to, and the discussions I had with the conference attendees, for the most part philosophers of science, made me discover various strands of the philosophical scholarship on causality that I had ignored, including the literature on Woodward's theory of "counterfactuals", Spirtes's work on automatized causal search algorithms, Guala's reflections on experiments and simulations, Reiss's theory of evidence, and Williamson and Russo's "evidential variety" thesis. On the other hand, at that conference, I met a philosopher, Lorenzo Casini, who, just after my keynote on "Agent-based models and types of causality in sociology", presented a paper on the use of agent-based models in finance. After discussing our respective talks, we decided to keep in contact with the possible aim of writing a common piece on what we discovered to be a common interest, i.e. the connections between agent-based models and causality.

For about two years Lorenzo and I struggled with our respective disciplinary habits and tried to find a compromise between my goal of being understandable by quantitative sociologists in order to try to modify the way many of them usually see causality through the specific lenses of multivariate statistics, and Lorenzo's wider philosophical ambition of theorizing the reliability of causal claims. Our reflection materialized in a conference paper that I presented at Harvard University in June 2015 in the closing panel of the eighth annual conference of the International Network of Analytical Sociologists. Interestingly, that round table was titled "Causal inference and mechanism-based explanation: friends or foes?" The feedback I received encouraged Lorenzo and I to develop the conference paper further, and we finally submitted a new version of it to a top journal in December 2015. The submission was unsuccessful. One referee was enthusiastic whereas the other engaged in a systematic critique of the paper's main thesis that agent-based modeling can play a central role in the causal inference business. The content and the style of the negative review made Lorenzo and me think that our argument was likely to hurt too directly the mainstream perspective on causal analysis among quantitative social scientists; a bit discouraged, we agreed to leave the paper aside for a while. Almost one year later, we decided to give the paper the status of a working paper and we were happy to see it accepted within the working paper series of the Institute for Analytical Sociology (see Casini and Manzo 2016).

I must thank Andrew Abbott who, in a private correspondence, intelligently persuaded me fully to forget about what the discontented reviewer had to say against the possible connection between agent-based modeling and causality, and motivated me to make the effort to develop the original paper into a book. Lorenzo agreed that I took the responsibility for this task, his interest having in the meantime evolved towards developing our initial argument on agent-based models and causality along more explicit philosophical lines, in particular within a formal Bayesian framework.

This essentially is the tortuous path through which I progressively came to prepare, between the fall of 2018 and the spring of 2019, a first version of the present book, which I initially submitted as part of my application for obtaining the *habilitation à diriger des recherches* (HDR), i.e. the post-doctoral degree that, in the French academic system, confers the right of supervising doctoral students and entitles one to apply for full professorship. The HDR defense took place on the July 6, 2019, at Sorbonne University, and it was another crucial step towards elaborating further my thoughts on agent-based modeling and causal inference. My work was indeed examined by a jury composed of Pierre Demeulenaere (my HDR sponsor at Sorbonne University), Anouck Barberousse, Ivan Ermakoff, Andreas Flache, Olivier Godechot, and Daniel Little. All of them wrote detailed reports on my manuscript, and, during the oral defense, they in a friendly way but rigorously identified various inconsistencies and points of my arguments that needed clarification and further elaboration. More or less at the same time, Christopher Winship also provided extensive written comments on the entire manuscript.

The book readers currently have in their hands resulted from my attempt to meditate on, assimilate, and respond to all the comments I received from these generous readers. I would like to express my deepest gratitude to all of them.

Alison Oliver, first, and, then, Kimberly Monroe-Hill at Wiley provided continuous support and patiently accepted my repeated delays. I enormously appreciated their assistance. I am indebted to Flaminio Squazzoni who, as one of the editors of the Computational and Quantitative Social Science Series at Wiley, accepted to consider my manuscript for publication as well as to three anonymous reviewers who initially evaluated my book proposal. I would also like to thank Lucas Sage for reading the entire manuscript and identifying various typos. I am obviously the only person responsible for the book's current content, and possible remaining omissions and mistakes.

Last but not least, I would like to thank my exceptional wife, Stéphane, for her daily love and patience, as well as my three splendid daughters, Eléonore, Mathilde, and Béatrice, for reminding me every day that academia is less important in life than spending time with them.

Paris, April 7, 2021

The Book in a Nutshell

Does grandsons' status depend on grandparents' socio-economic position (Chan and Bolliver 2013)? Do highly religious, educated, and economically active Muslim women veil more than Muslim women of comparable religiosity but with lower exposure to urban and modernized settings (Aksoy and Gambetta 2016)? Does the mandatory character of teachers' recommendations increase the inequality in educational outcomes across socio-economic and ethnic groups (Dollmann 2016)? Does prisoners' social status mainly depend on their age and the time they spent in prison (Kreager et al. 2017)? Does multiple organization membership reduce the probability that individuals adopt certain behaviors (Brashears et al. 2017)? Does voluntary inter-firm mobility have an impact on mobile individuals' wage, and, if so, does gender modify this dependence (Pearlman 2018)? Does dis-unionization contribute to the rise of wage inequality (VanHeuvelen 2018)? Do actors with low social status disproportionately attract sanctioning and repulsive acts (Rubineau et al. 2019)? Does educational attainment depend on cognitive abilities and, if so, does the socio-economic status of the pupils' family modify this connection (Stienstra et al. 2020)? Does job loss increase the probability of union dissolution (Anderson et al. 2021)? Does individuals' education affect individuals' wage over time (Cheng 2021)?

Despite the different substantive focus, these questions all deal with dependence relationships between events, states, or features of the social world and of individuals populating it. The ultimate goal of each study is to establish the extent to which the variability across the units of analysis of interest on a given property makes a difference for the value of another given property of these units. In this sense, these questions all are causal questions. The most visible European and North American scientific journals in sociology are overwhelmed by this type of question, which is typically answered through sophisticated statistical methods or ingenious experimental designs. This type of research is consequential beyond the realm of quantitative and experimental social sciences. Ethnography-oriented works published in the most prestigious sociological journals indeed also tend to be informed by quantitative causal reasoning and language (see Abend et al. 2013).

In this book, I question the way empirical and experimental social research in sociology typically build causal reasoning. I want to defend the view that the persuasiveness of causal claims depends on the co-presence of two conditions: first, conditionally on certain theoretical assumptions, a robust dependence relationship must be proven to exist empirically between two happenings; *and*, second, a plausible (meaning, theoretically and empirically realistic) model of a mechanism that is likely to generate this relationship must also be provided. I will argue that this view is likely to arouse skepticism among scholars involved in causal analysis in sociology because their typical recipe to establish causal claims is based on two generic assumptions that impede seeing that causal inference requires complementarity between various types of information and different methods of analysis.

The first of these two assumptions is that causal inference is *essentially* believed to be a matter of empirical data. This does not mean that quantitative scholars interested in causal analysis have a naïve view of causal inference. The best articles typically contain explicit warnings about the difficulty of establishing causality from even robust correlational evidence and/or exogenous manipulations (for typical formulations, see Aksoy and Gambetta 2016: 803); the most competent scholars regularly discover identification and specification problems that went unnoticed in previous studies (for a recent case, see Breen 2018); papers published in high-quality journals clearly must show awareness of, and propose possible solutions to, omitted variable problems and reverse causality issues (for an example, see Aven and Hillmann 2017: 8–10). Actually, what is typically absent from the published articles is an explicit discussion of the modeling assumptions that are hard or even impossible to test with data but are needed to make the causal claim hold. This is a problem with a fractal nature. It arises with the first set of data and methods used, and it arises again and again for each technical solution one may implement to solve the problems left unsolved by the previous set of analyses. Judea Pearl (2009: 100, italics added) expressed this point clearly when he posed the following "*golden rule*": "behind any causal conclusion there must be some causal assumption, untested in observational studies." This is precisely what is systematically not problematized in current causal research in sociology. This is consequential: in the end, causality is thought of as *ultimately* being a matter of empirical data.

Throughout this book, I instead restlessly emphasize the idea that causality is always a matter of both empirical data and theoretical assumptions (be they formal or substantive) that one must continuously build to justify those assumptions for which data are absent in practice and/or cannot be produced in principle. To defend this first claim, I will scrutinize a specific type of numeric simulations, in particular agent-based computational models that are *a priori* excluded from the causal enterprise by the vast majority of quantitative scholars, and compare them to certain statistical methods and to experiments. I will argue that untested, and untestable, assumptions are in fact present in both observational/experimental and simulation methods.

The set of articles that I mentioned in the first paragraph illustrates a second widespread implicit assumption of quantitative causal reasoning in sociology: causal inference is *ultimately* thought to be possible without making explicit a *formal substantive model* of a mechanism potentially generating the dependence relationship of interest. Substantive models may come in two different forms. On the one hand, one may formulate specific hypotheses on the actions, interactions, and social constraints that may have generated a given variation in a given outcome across a given set of units of analysis, and use this formalization to give a specific theoretical meaning to the parameters of the estimated statistical models aimed at establishing the causal connection. This corresponds to what Skvoretz (2016) called *theoretical models*—as opposed to "methodological models", i.e.

models designed to discover relations in the data. On the other hand, one may design a dedicated numerical device that, on the basis of specific hypotheses on actions, interactions, and social constraints, generates numeric data with the aim of seeing whether the supposed causal connections observed in the empirical data also appear in the data for which we know the production mechanism. This corresponds to the logic of what Boudon (1979) called *generative models*.

Early examples of causal analysis based on *theoretical models* can be found in Sorensen's (1977) study of the role of vacancy chains in the generation of income inequality, and more recent illustrations are Skvoretz's analysis of homogamy (see Skvoretz 2013; Karpiński and Skvoretz 2015) or Kruse's (2017) investigation of interethnic friendship formation among German adolescents. Experimentalists also rely sometimes on mathematical (game-theoretic) models in order to make predictions to be checked in the laboratory. Recent illustrations of this strategy are Becker et al.'s (2017) study of the impact of social influence on the accuracy of collective judgments, and Frey and van de Rijt's (2016) analysis of the effect of reputation accumulation on the amount of inequality in transaction volumes among sellers in market exchanges. But, again, many other field and laboratory experiments only generically refer to a given theory (or set of ideas) that may justify the presence of supposed causal effects (see, for instance, Melamed and Savage 2016 or Grossman and Baldassarri 2012). *Generative models* are even rarer among quantitative social scientists interested in causal inference. An interesting exception is Bruch's (2014) study of the effect of income inequality between (and within) racial groups on racial residential segregation where an explicit focus on establishing a causal connection through statistical methods is coupled with various generative models implemented in empirically calibrated agent-based computational simulations. Similarly, Manzo et al. (2018) backed the observation of variations in the causal effect of family networks on the diffusion of innovations between religious groups with empirically grounded simulated generative models.

These are exceptions, however. The general tendency is to exclude *formal substantive models*, be they theoretical or generative, from quantitative causal reasoning. Causation is thus implicitly seen as a matter of robust correlation or consequential manipulation—or, if one prefers, as a choice between "reverse" and "forward" causal inference (see Gelman 2011). In the vast majority of statistically oriented and experimental works looking for causal effects there is simply no room for a different understanding of causality that could complement these two general perspectives on causality.

Throughout this book, I instead restlessly emphasize the idea that in fact another understanding of causality exists that makes the formulation, and the empirical test, of *substantive models* (i.e. formal models of a mechanism) a necessary condition for persuasive causal claims. The intuition behind this view is that, without empirically supported substantive models, doubts about the possible presence of confounders cannot be dissipated. In particular, I will argue that a specific type of numeric simulations, namely agent-based computational models, is especially flexible to design substantive models of a generative kind. The bulk of my contribution can then been found in the analysis that I will propose of the conditions under which agent-based computational models can lead to theoretically and empirically realistic substantive models, thus contributing to the construction of persuasive causal claims.

Thus, in a nutshell, this book wants to defend the following two intertwined theses: first, a causal claim should always be thought of as a matter of empirical data, *plus* theoretical arguments that are needed to defend untested and untestable assumptions of the chosen method to describe/produce those data; and, second, various, but complementary, understandings of causality exist and are supported by different, but complementary, methods. The persuasiveness of a given causal claim, I will argue, increases proportionally to our capacity of combining methods that are designed to describe dependence relationships and methods that are conceived to design formal models of generative mechanisms, and by paying due attention to the fact that both types of methods rely on a complex combination of empirical data and theoretical arguments.

To defend this perspective, the book builds on theoretical resources that already exist within the sociological, philosophical, and methodological literature on causality but that, in my view, have not been fully exploited yet to think about causal reasoning in quantitative sociology.

Among these resources, let me first emphasize a theory of causality—namely, the "epistemic theory of causality" (see Williamson 2006)—that invites us to conceive causal analysis as the dialogic exchange between the investigator and a given audience, where the investigator has the responsibility of producing empirical facts and theoretical arguments potentially reducing the uncertainty the audience may feel as to the persuasiveness of the causal claim at hand. The specific point of the epistemic theory of causality that I exploit in this book is that, according to this approach, for an observer to develop rational causal beliefs, several pieces of evidence, qualitatively different (i.e. arising from different points of view of the world), should be conveyed to her. This property is called "evidential variety" or "evidential pluralism" and it is regarded as having a crucial impact on the degree of trustworthiness one can develop regarding a given causal claim (see Russo and Williamson 2007). Another important theoretical resource on which I build part of my analysis is the variety of definitions of the concept of causality that the philosophical literature has progressively elaborated (see Hall 2004), which makes doubtful the attempt of many quantitative social scientists to restrain the concept of causality to a certain understanding of causality. Similarly, the concept of mechanism in fact receives different treatments within quantitative social sciences (see Kalter and Kroneberg 2014), which, again, pushes to be pluralistic.

Thus, this book can be seen as an attempt to map the existence of a variety of points of view that already exist in the literature (see Chapter 1) and combine them into a coherent and comprehensive framework for causal inference in quantitative social sciences based on the "evidential variety" thesis (see Chapter 6). This goal echoes recent calls for methodological pluralism from scholars who travel across disciplines. In particular, Page (2018: 1) proposed to invest more effort in the "many-model thinking approach", i.e. "the application of ensembles of models to make sense of complex phenomena". His justification for this approach is that different models, focusing on "distinct causal forces" and "different scales" produce "wisdom through a multiplicity of lenses". Page (2018: 5) explicitly opposed the many-model approach to the traditional approach, where "the objective is to (a) identify the one proper model and (b) apply it correctly."

In this book, I follow the same logic with respect to the purpose of making persuasive causal claims. I want to show that the typical logic "one causal question; one understanding of causality; one proper method" is insufficient. We should accept instead that establishing a persuasive causal claim is a complex task that requires several sources of empirical evidence and arguments as well as the synergy of several methods.

To those who would be tempted to object that this view in fact equates "causal inference with scientific explanation (...)"—"(...) instead of the standard use for confrontation of causal statements with empirical observations", to borrow the words of a reviewer I met along the way—, I would retort that the objection precisely arises from the two implicit assumptions that I have described above, namely the widespread conviction that causality ultimately is a matter of empirical data and that formal models of generative mechanisms simulating the potential sources of the connection of interest are not part of the causal business. With this book, my hope is that I can contribute to demonstrating that this supposedly "standard" view ignores entire strands of scholarship showing that both assumptions are unnecessary ingredients that make causal inference weaker, not stronger.

Introduction[1]

The concept of causality and quantitative techniques for causal inference have been extensively discussed in sociology by a variety of scholars with different research programs and theoretical perspectives (see, among others, Marini and Singer 1988; Abbott 1998; Doreian 1999; Goldthorpe 2001; Winship and Sobel 2004; Mahoney 2008; Gangl 2010; Mahoney et al. 2013; for a historical perspective, see also Barringer et al. 2013). Among quantitative sociologists, the interest in causality issues was reinforced by the rapid diffusion in sociology of the potential outcome approach to causality (Morgan and Winship 2015), an approach in turn fostered by older contributions in statistics (for a historical overview, see Imbens and Rubin 2015: ch. 2) and economics (Heckman 2005), and reinvigorated by recent developments in computer science (Spirtes et al. 2000; Pearl 2009) as well as philosophical discussions (Woodward 2003). Extensive literature reviews documented that establishing causal claims is now one of the primary goals of many articles published in leading sociological journals, be these quantitative, qualitative, or historical studies (see Abend et al. 2013; Ermakoff 2019).

Traces of the concept of causality can also be found in the literature on mechanism-based thinking (Hedström and Ylikoski 2010) and on agent-based computational models (Bianchi and Squazzoni 2015). And these are also two research topics in rapid expansion in contemporary sociology.

As to mechanism-based explanations, it is true that the effort towards identifying generalizable fine-grained chains of small-scale events with clearly defined large-scale consequences can be traced back to the infancy of modern social sciences (see, for instance, Elster's 2009b study of Tocqueville's œuvre). It is also true that the notions of "social mechanism" and "generative model" were initially forged by mathematical sociologists in the 1960s (see, respectively, Karlsson 1958: 16; Fararo 1969a: 225, 1969b: 81, 84–5; see also Boudon 1979). At the same time, it seems correct to regard Hedström and Swedberg's (1998) collection on social mechanisms, coupled with philosophical studies of research practices in biology and neuroscience (Machamer et al. 2000; Bechtel and Abrahamsen 2005; Craver 2007), as the starting point of systematic investigations on the concept of mechanism-based explanation. As a by-product, analytical sociology has progressively emerged as a distinctive style of social inquiry (see, among others, Hedström 2005;

---

[1] This introductory chapter builds on and extends Casini and Manzo (2016: 2–6).

Hedström and Bearman 2009; Demeulenaere 2011a; Manzo 2014a, 2021). Discontents with analytical sociology, in turn, argued that mechanism-based explanations can in fact be framed in different ways (see, among others, Abbott 2007; Gorski 2009; Gross 2009; Sampson 2011; Little 2012; Opp 2013; for a reply, see Manzo 2010, 2014a).

As to agent-based computational models, hereafter ABMs (or ABM, for the singular form and for "agent-based modeling"), the basic principles appeared in pioneering studies in the 1960s (Hägerstrand 1965; Sakoda 1971; Schelling 1971) but their diffusion accelerated after the publication of systematic monographs such as Axtell and Epstein (1996), Axelrod (1997), and Epstein (2006). Nowadays, pleas for ABMs exist in a large variety of disciplines—including biology (Thorne et al. 2007; Chavali et al. 2008), ecology (Grimm et al. 2006), macroeconomics (Farmer and Foley 2009; De Grauwe 2010), quantitative finance (Mathieu et al. 2005), organization and marketing studies (Fioretti 2013), political science (Cederman 2005; de Marchi and Page 2014), geography (O'Sullivan 2008), criminology (Birks et al. 2012), epidemiology (Auchincloss and Diez Roux 2008), social psychology (Smith and Conrey 2007), demography (Billari and Prskawetz 2003) and archeology (Wurzer et al. 2015). Sociology is no exception (Macy and Flache 2009). Leading journals have started paying attention to ABMs (Gilbert and Abbott 2005; Hedström and Manzo 2015) and the number of applications at the core of the discipline is fast increasing (Macy and Willer 2002; Sawyer 2003; Bianchi and Squazzoni 2015).

The starting point of this book is that, although scholarship on causality, mechanisms, and ABMs is burgeoning, a systematic discussion of the conceptual and methodological connections between these three topics is still missing. A few examples suffice to document this fact.

Among social scientists, Hedström and Ylikoski (2010) reflect on the concept of both cause and mechanism, but, when they treat ABMs, the issue of the potential contribution of this method to causal inference is not addressed. Demeulenaere (2011b: 12–20) explicitly studies the connections between the concepts of mechanism and causality—making the important point that it is in fact disputable to regard the former as a substitute for the latter because, he argues, any mechanism, to work as such, must rely on causal regularities, a general argument that he also applies to the more specific problem of the explanations of individual actions (see Demeulenaere 2011c). The implications of this argument for methods for causal inference are not addressed, however. Knight and Winship (2013) criticize the way the concept of mechanism is employed within the analytical sociology literature; they propose a more precise definition of the concept, which they regard as compatible with a counterfactual view of causation, and show how this definition can be employed, using directed acyclic graphs, to identify causal relations; computational methods, however, receive no mention. Watts (2014) reflects on the notion of causal explanation in connection with a critical analysis of a specific aspect of the mechanism-based perspective, namely action theory, but, when he addresses the methodological side of the issue, experimental and statistical methods are only quickly discussed and no attention is devoted to ABMs. Finally, let me mention Gross (2018) who has interestingly argued that mechanism-based explanation with causal ambitions should pay more attention to the formal properties of causal chains but, once again, how this translates, on a methodological level, in specific procedures for causal inference is not discussed.

Philosophical investigations exhibit a similar pattern. Several articles scrutinize the connection between the concepts of causal and mechanistic explanation (Glennan 1996, 2002; Woodward 2002, 2013; Casini et al. 2011; Menzies 2012; Williamson 2013); however, the discussion of what techniques would support the connection between methods for causal inference and strategies for mechanistic explanation is limited, and, in any case, either does not contemplate ABMs at all (for some illustrations, see Steel 2004; Reiss 2009; Mouchart and Russo 2011; Hoover 2012) or only points to this technique without any detailed analysis (see, for instance, Kaidesoja 2021a, b). Analogously, the rare philosophical contributions specifically discussing the potential relevance of ABMs for supporting causal explanations—some denying such relevance (Grüne-Yanoff 2009a), others arguing in favor of it (Elsenbroich 2012; Casini 2014)—only rely on specific models, lack a systematic discussion of theories of causal inference and social mechanisms, and never systematically confront ABMs with other methods for causal inference (for an example of the latter limitation, see Anzola 2020).

In this book, my intention is to show that a comprehensive analysis of causality, mechanisms, and ABMs *at the same time* has the potential of modifying our understanding of causal inference within quantitative social sciences.

1 The Book's Question

The absence of systematic discussions on the possible three-way connections between causality, mechanisms, and ABMs left unsolved an important question: *in what sense* and *under which conditions*, if any, can ABMs contribute to causal inference? This book has the ambition to provide a principled answer to this question by scrutinizing, and systematically connecting, scholarship on (methods for) causal inference, social mechanisms, and ABMs

Answering this question is complicated by the variety of views that exist on the status of an ABM itself. If many social scientists (see, among others, Hummon and Fararo 1995b; Epstein 1999, 2006: chs. 1–2; Axtell et al. 2002; Sawyer 2004; Cederman 2005; Tesfatsion 2006; Manzo 2014a) and philosophers of social science (Ylikoski and Marchionni 2013: §3) agree on ABM's ability to act as a methodological lever for mechanism-based explanations, disagreement remains on the extent to which an ABM can also be used to persuade an audience that the postulated mechanism is operating in the real world.

Let us first consider philosophers' views on this point. Grüne-Yanoff (2009a) studied a famous, empirically grounded ABM of a particular historical phenomenon (the so-called artificial Anasazi model by Jeffrey Dean and collaborators) and, on this specific basis, generally denied that this method can help causal inference. On the one hand, he argued, "full" causal explanations cannot be provided by an ABM because it is not possible to obtain empirical evidence on all the elements constituting the mechanisms designed by the model; on the other hand, Grüne-Yanoff continues, even potential explanations, meaning "possible causal histories", cannot be supported by an ABM because the method lacks an internal criterion to select among all the possible options.

Casini (2014) defends a different view. He scrutinized two specific highly stylized ABMs aiming at the generation of stylized macroscopic targets, and argued that even hyper-simplified ABMs can in fact produce potential explanations that increase our understanding of the mechanism at work, thus contributing to identifying actual causal forces. This is possible, Casini (2014: 665) claimed, as long as the model is proved to be "credible-with-respect-to-the target", which, according to him, depends on the following three conditions: "(i) the soundness of theoretical principles, psychological assumptions, and functional analogies; (ii) the robustness of the results across changes in initial conditions and parameter values; and (iii) the robustness across changes in modeling assumptions." According to Casini, robustness analysis is especially important because this operation allows it to be shown that "the mechanisms represented by the models are different 'tokens' of the same type" (*ibid.*: 666), which suggests that the model at hand captures realistic, not accidental or artificial, features of the actual causal story.

Through a different language, Ylikoski and Aydinonat (2014) defended a similar argument. By studying a very abstract ABM, they argued that systematic variations of the initial model's assumptions (i.e. robustness analysis) lead to a "clusters of models". These clusters can contribute to causal analysis in two ways. One the one hand, they help us to see what is essential in the original model, thus helping to formulate hypotheses on the "core" features of the actual causal stories; on the other hand, clusters of models help us to formulate alternative explanations, which allow us to better locate a given causal story to be empirically tested within the set of possible alternative causal stories.

These philosophical analyses led to different conclusions as to the value of ABMs for causal reasoning but they shared an implicit assumption: as long as an ABM cannot generate *on its own* all the required empirical data to corroborate the assumptions on which it was built, the ABM can *at best* guide, or complement, methods for causal inference relying on empirical data but it cannot ascertain *on its own* actual causal stories. However, as noted by Elsenbroich (2012) in her response to Grüne-Yanoff's critique, "(...) insisting on complete knowledge of microphenomena for a causal explanation makes causal explanation impossible in the social sciences. This is also not a problem of ABM but of social science as a whole." I will develop this important point further when I discuss data-driven methods for causal inference and show that these methods, too, can make causal claims only contingently on assumptions that they cannot test empirically *on their own* (see Chapter 5).

Social scientists formulated similar views on the basis of similar implicit assumptions. Macy and Sato (2008), in response to the critique of lack of realism of an ABM that they proposed, defended their model by claiming that "[t]he computational model can generate hypotheses for empirical testing, but it cannot 'bear the burden of proof'", thus implicitly proposing a division of labor according to which the ABM is a tool for theoretical exploration while experimental and statistical methods for observational data are better suited, and necessary, to support causal inference. Quantitative scholars relying on this type of method overtly expressed skepticism about "(...) the utility of many simulation-based methods of theory construction" (Morgan and Winship 2015: 341). As clearly visible in Morgan's (2013) overview of the most recent developments in the field of causal inference, the ABM is simply not considered as a potential player in this game.

However, other scholars have noted that an ABM can communicate with empirical data in various ways, which in principle makes it capable of pinpointing real-world mechanisms underlying the dependence relationship between variables (Hedström 2005: ch. 6;

Manzo 2007; Bruch and Atwell 2015). This view has again been recently attacked by Törnberg (2019) who argued that, because of the contingent nature of the social realm, it is impossible to know how a simulated mechanism will behave in the real world. For this reason, he claimed, "whether these mechanisms are then empirically manifested or generate a certain effect in reality is a matter for empirical investigation to conclude" (*ibid.*: 9). As a consequence, Törnberg regarded "(...) the idea of fitting the model to empirical data—in other words, to carefully validate or calibrate models by matching regularities produced in the model with empirical regularities—(...)" as "fundamentally misleading", an idea that, according to him, "stems from a confusion of abstract categories and concrete phenomena" (*ibid.*: 11). For this reason, Törnberg concluded that ABMs should only be seen "as a form of computer-aided abstraction that boosts our intuition and thus helps us to disentangle such complex webs of causality" (*ibid.*: 13).

Interestingly, the same opposition appeared outside sociology. In epidemiology, Marshall and Galea (2015) provided a systematic analysis of ABMs from the point of view of counterfactual reasoning, and argued that, to the extent to which an ABM can be seen as an implementation of the potential outcome approach to causal inference, an ABM can support causal inference (see also Anzola 2020: 55). To this, Diez Roux (2015: 101) has replied in the following way:

> (...) there is a fundamental distinction between causal inference based on observations (as in traditional epidemiology) and causal inference based on simulation modeling. The traditional tools of epidemiology are used to extract (hopefully) reasonable conclusions from necessarily partial and incomplete (often messy) observations of the real world. (...) In contrast, when we use the tools of complex systems, we create a virtual world (based on prior knowledge or intuition) and then explore hypotheses about causes under the assumptions encoded in this virtual world. In the simulation model, we cannot directly determine whether X causes Y in the real world (because the world in which we are working is of our own creation); we can only explore the plausible implications of changing X on levels of Y under the conditions encoded in the model. In the real world, we have fact (what we observe) and we try to infer the counterfactual condition (what we would have observed if the treatment had been different). In the simulated world, everything is counterfactual in the sense that the world and all possible scenarios are artificially created by the scientist.

Diez Roux's statement is the clearest illustration I could find in print of the widespread belief that statistical and experimental methods that now have received a unified counterfactual interpretation within the potential outcome framework (see Imbens and Rubin 2015; in sociology, Morgan and Winship 2015) are intrinsically different from, and superior to, ABMs with respect to their capacity to produce persuasive causal claims.

In this book, I want to suggest that these contrasting views on the *sense* and *the conditions* (if any) under which ABMs can contribute to causal inference are limited by the specific way the question was addressed. In particular, existing analyses share the three following features: (i) they stem from a specific understanding of the concept of causality *while* a systematic analysis of the philosophical literature on this concept shows that causality can in fact be understood in very different ways; (ii) they move from specific views on the concept of mechanism *without* considering that this concept can in fact be understood in different ways, a variety of definitions that interestingly square with different approaches to causality; and (iii) existing analyses never systematically compare ABMs with observational and experimental methods for causal inference with respect to the balance between empirical data and theoretical assumptions on which these methods are built.

With this book, my goal is to overcome these three conceptual biases and show that, when scholarship on traditional methods for causal inference, social mechanisms, and ABMs are systematically scrutinized, a comprehensive view emerges where ABMs play a pivotal role for causal inference in quantitative social science.

2 The Book's Structure

To develop the argument, I go through two steps. They correspond to the two parts in which the book is organized. Each part in turn contains three chapters.

In the book's Part I, I make an effort of conceptual and methodological clarification that is needed to properly address my central question, i.e. *in what sense* and *under which conditions* ABMs can contribute to causal inference.

In particular, Chapter 1 treats the notions of causality and mechanism, and stresses the diversity of accounts that both concepts have received. My goal is to establish that claims about methods' potential for causal inference as well as about how to model social mechanisms implicitly rely on one's specific understanding of the notions of causality and mechanism. As ABM is especially compatible with one specific understanding of the concepts of causality and mechanism— it follows that some see ABMs' potential for causal reasoning while others do not.

Chapter 2 moves from conceptual to methodological clarifications. It discusses generic features of ABMs, and clarifies the technical reasons behind the affinity of this method with one specific understanding of the concept of causality and mechanism. Thus, what was only posed in Chapter 1 is explained in Chapter 2. In this chapter, I also discuss the role and the meaning of "variables" in ABMs, an important point that, if properly understood, may favor a better dialogue between computational and statistically minded modelers.

Chapter 3 deepens the methodological analysis of ABMs, and adds an important clarification. In particular, I stress that existing ABMs are various with respect to (i) the theoretical plausibility of the assumptions that are used to design the model (which I will call "theoretical realism"), (ii) the type of information that is used to build it (which I will call "empirical calibration"), and (iii) the nature of the data that are used to assess the validity of the model's implications (which I will call "empirical validation"). To map this diversity, I analyze typical ABMs that contributed to establish the ABM field as well as a selection of recent ABMs from a variety of disciplines. On the basis of this meta-analysis, I argue that, since all three dimensions impact on an ABM to produce reliable results, a principled assessment of the usefulness of ABMs for causal reasoning cannot neglect the variety of available ABMs, which means that any conclusion drawn from studying this or that specific ABM will have a limited scope.

With such conceptual and methodological qualifications in place, the book's Part II addresses my central question, i.e. *in what sense* and *under which conditions* ABMs can contribute to causal inference.

In particular, Chapter 4 discusses the *in principle* conditions for ABMs to contribute to causal inference, and relates them to the three dimensions (i.e. "theoretical realism", empirical "calibration", and "validation"), discussed in Chapter 3, along which ABMs can be classified. I then highlight the *in practice* obstacles to the realization of these *in principle* conditions, and describe the research operations that ABM modelers can perform when these *in principle* conditions for causal inference cannot be satisfied in practice.

Chapter 5 plays a crucial role within the structure of the argument. By focusing on a set of methods that are usually regarded as optimally suited to establish causal relationships, namely randomized control trials (RCTs), instrumental variables (IVs), and causal graphs (in particular, directed acyclic graphs, DAGs), I show that, on close scrutiny, these methods, when they are used for causal inference, face challenges that are identical to the obstacles faced by ABMs seeking realistic mechanism-based explanations. Similar to the latter, I argue, well-established methods for causal inference in fact also face issues of data availability, untestable assumptions, and uncertain reliability.

Chapter 6 draws all the implications from the observation that, similarly to ABMs, traditional methods for causal inference support causal claims only if specific, and highly demanding, conditions are fulfilled. In particular, I call for shifting the focus of the debate from proposing arguments that defend the superiority of particular methods on the basis of the (often implicit) endorsement of particular notions of causality and mechanism to discussing how different methods may coexist and ought to cooperate. To motivate this call, I finally explain why different methods produce different kinds of evidence and theoretical arguments, and argue that, since in practice every kind of evidence and theoretical argument is likely to be imperfect, the persuasiveness of causal claims can only be built at the intersection of these methods, the weakness of one method being compensated by the strength of another.

In (the Coda), I first examine potential objections to my main claims and then provide a detailed summary of the analysis.

Part 1

Conceptual and Methodological Clarifications

1

The Diversity of Views on Causality and Mechanisms

This chapter is devoted to three conceptual clarifications. First, I clarify the meaning I give the concept of causal inference throughout the book (Section 1.1); second, I discuss the concept of causality (Section 1.2); and, finally, I investigate the concept of mechanism (Section 1.3). An explicit discussion of the variety of meanings the concepts of causality and mechanism usually receive in different academic communities is useful to highlight an interesting fact: specific views on causality tend to square with specific views on mechanisms. Once this is clearly seen, it will become apparent that the observed disagreements on the usefulness of an agent-based model (ABM) for causal inference ultimately arise from the fact that scholars in different methodological traditions endorse conflicting views on what establishing causality and identifying mechanisms mean (Section 1.4). To establish this fact is the first step, I believe, to building a pluralistic and comprehensive view on causal inference in quantitative social sciences.[1]

1.1 Causal Inference

Historically, statistical inference was created to solve the fundamental problem of gaining knowledge about given features, i.e. parameters, of a given population by observing only a portion of it, i.e. the sample (see Goldthorpe 2021: chs. 3–4)? Probability theory—in its frequentist and subjective variants (see Cox 2006)—was proposed as the tool for quantification of the errors that one inevitably makes when studying the parts rather than the whole.

Once this mathematical invention entered social sciences, statistical inference was stretched in two different, but related, ways. On the one hand, as described by Ziliak and McCloskey (2008), social scientists have progressively tended to give priority to establish whether parameters, and their association, are different from zero rather than focusing on estimation of their value. Thus, the question of the existence of an effect, rather than its size, becomes the main goal of statistical inference, and statistical tests, with their significance values, the main tool to make a decision. On the other hand, as noted by Freedman (1995), many social scientists were incapable of resisting the temptation of considering a presumably different-from-zero association between two features as the sign of the existence of a

---

[1] This chapter builds on and extends Casini and Manzo (2016: 6–14).

causal connection between them. Thus, various forms of regression analysis became the main tools to establish causal claim, and statistical inference was silently equated to causal inference. In this way, Freedman (2005) notes, a second interpretative layer was added. The same set of assumptions that was initially created to determine whether, under a specific theory of (a specific type of) errors, a feature had a certain value in the population was now extended to the interpretative, not factual, task to make this value contingent on the estimated value of another feature. The shift that was progressively operated in quantitative social science from statistical inference to causal inference is nowadays especially visible within the literature on the potential outcome approach as a comprehensive framework to interpret both experiments and multivariate statistics for observational data (see, in sociology, Morgan and Winship 2015; in economics, Imbens and Rubin 2015).

Throughout this book, I adopt a generic definition of causal inference, which I regard as the cognitive operation that consists in using fragmentary empirical and theoretical information to establish the extent to which one specific happening systematically alters the probability that another happening follows. This definition is motivated by the theory of "epistemic causality" (see Williamson 2006) and the "evidential variety" thesis (see Russo and Williamson 2007), which consider that defending a causal claim always requires an investigator and an audience: the investigator's goal is to accumulate empirical data and theoretical arguments that must be as diverse as possible to make her causal claim as persuasive as possible in the eyes of the given audience. From this perspective, it is unlikely that the investigator is able to persuade the audience that her causal claim is plausible as long as she only produces a single type of evidence originating from a single type of method. As a consequence, there is no reason to restrict *ab initio* the concept of causal inference to a specific type of method and evidence, namely statistical methods for survey data or experimental designs. This open view of what counts as evidence for causal inference echoes some statisticians' call for recognizing that confidence in statistical methods and statistical results is always built in practice at the intersection of a complex mix of elements of different natures, which suggests problematizing what "convincing evidence" is in practice (see, in particular, Gelman and O'Rourke 2013; Gelman and Basbøll 2014).

Consistently with this view, let me emphasize that, although I will argue that generating the connection of interest from formal models of mechanisms through ABMs constitutes the "severe test" for making persuasive causal claims (on "severe testing", see Mayo 2018), I restrain myself from defining causal inference in general in terms of mechanism-based reasoning (for a different choice, see Hedström 2009). In my view, distinguishing analytically the cognitive operation of inferring causality from the specific types of arguments and methods one exploits to perform this operation is the very preliminary conceptual step we need to appreciate precisely that a variety of methods, data sources, and theoretical arguments actually should be combined to make persuasive causal claims.[2]

---

[2] This seems the right point of the analysis to make an important linguistic observation. From the very beginning of the book, I tried to be explicit about one of the major points I want to make, namely the idea that assumptions are a crucial part of any causal reasoning, which implies that "evidence" can never be conclusive *alone* because it is always conditional on those assumptions. Despite this starting point, in a previous version of the manuscript I recurrently adopted the expressions "convincing evidence" and "convincing causal claims". Christopher Winship made me realize that this linguistic choice may have

1.2   *Dependence* and *Production* Accounts of Causality

To argue in favor of a pluralistic perspective on causal inference, the first step is fully to acknowledge that it is hard to restrict the concept of cause to a single, and uncontroversial, meaning. The point is nicely illustrated by Nancy Cartwright (2004: 806, italics mine), when she notes that "[t]he term cause is highly unspecific. It commits us to nothing about the kind of causality involved nor about how the causes operate. Recognizing this should make us *more cautious* about investing in the quest for *universal methods for causal inference.*"

This is an important epistemological observation that has the following implication. If the concept of causality can theoretically be understood in different ways, and if different quantitative methods are differently permeable to different theories of causality, then it is likely that one's view on what method is best suited to perform causal inference in fact depends on one's favored intuitions on what causality is or is not. Thus, making explicit such intuitions is especially important for assessing in what sense and under which conditions methods that are usually excluded from the causal business—like ABM—can contribute to causal inference *compared* to more well-established experimental and observational methods.

Contrary to the skepticism of many quantitative social scientists against philosophers of causality[3] I do believe that philosophical writings on causation provide theoretical coordinates to map the different ways we can conceive of a causal relationship. In particular, for my purpose, the most relevant point is the generic distinction between *dependence* (or *difference-making*) accounts of causality and *production* accounts of causality (see Hall 2004). Roughly, among dependence accounts, one finds regularity, probabilistic and counterfactual views of causality whereas, among production accounts, one finds process, entities-andactivities and dispositionalist theories (for similar categorizations, see Kistler 2002; Psillos 2007; Reiss 2013: ch. 5). Obviously subtle differences between each variant of the two perspectives exist. For my argument, however, what matters most are the basic intuitions that inspire these two groups of theories of causation.

In particular, the idea behind dependence accounts is that causes are such that their obtaining makes a difference to the obtaining of their effects. In contrast, the idea behind

---

run against my own central argument because the adjective "convincing" may convey the impression that, despite all my qualifications, there can be in some sense some forms of absolutely convincing evidence (or at least that I believe that this is ultimately possible). To avoid this misunderstanding, I substituted "convincing" for "persuasive" (I must thank Winship again for suggesting this alternative), and limited the use of the first term to statements where no ambiguity seemed possible to me. For the same reason, I will very often write "data and arguments" rather than "evidence" with the intent of being entirely explicit about the point that the kind of support that any method can provide for a given causal claim is always a combination of partial empirical information and arguments that are necessary to defend assumptions that are not, or are only partially, empirically verifiable.

[3] For instance, Freese and Kevern (2013: 28) claimed: "The professional philosophical literature on causality is often surprisingly unhelpful: the practically-minded researcher digs in looking for clarity and instead is soon invited to consider examples of simultaneous assassination attempts or billiards rolled into time machines. No uncontroversial general philosophical account of causality exists, and social researchers have plenty of our own work to do while we wait".

production accounts is that causes are such that they help generate, or bring about, their effects. To Hall (2004: 226), from whom I am taking this conceptual distinction, dependence and production accounts are irreducible to one another, so we have distinct concepts of cause. In the present investigation, I remain agnostic on what causality is ontologically. My ultimate goal is to propose a comprehensive and pluralistic framework for causal inference that can accommodate several types of methods, sources of data, and theoretical arguments. For this reason, I am reluctant to accept that distinct and irreducible notions of cause exist. On a methodological level, this view has undesirable consequences, which I will discuss later (see Chapter 6).

For now, let me emphasize that, independently from philosophers of science, and under different terminological labels, sociologists have developed categorizations that follow Hall's distinction between dependence and production accounts of causality. In particular, more explicitly than others, Goldthorpe (2001) remarked that causation can be interpreted at least in three different ways as "robust dependence", as "consequential manipulation", or as "generative process". According to him, in the first case, "the causal claim depends on showing that X continues to affect Y when a set Z of other variables, also possibly related to Y, are introduced in the analysis" (*ibid.*: 2). In the second case, "genuine causation is that if a causal factor, X, is manipulated, then, given appropriate controls, a systematic effect is produced on the response variable, Y" (*ibid.*: 4–5). Finally, when causation is understood as a "generative process", "(...) what is important is the nature and the validity of the account given of the process that underlies the association appealed to (...)" (*ibid.*: 9). In terms of the aforementioned philosophical categories, causation as "robust dependence" and "consequential manipulation" clearly exemplify *dependence* accounts of causality, whereas what Goldthorpe labels causality as "generative process" naturally falls within *production* accounts.

Goldthorpe also observed that these views on causality combine in practice with distinctive methods of social inquiry. According to him, the view that causality essentially depends on controlling for confounders has informed time-series analysis, the early generation of causal path analysis, structural equations models, and more generally the large panoply of multivariate quantitative, regression-like techniques for survey data analysis (on this point, see also Abbott 1998; Freedman 2005). The "consequential manipulation" view squares with the methodology of randomized experiments. Interestingly, Goldthorpe hesitates to identify a specific method that illustrates the view of causality as a "generative process". Although he sees the potential of simulation methods as a possible option to test the validity of a proposed account of the underlying process (*ibid.*: 14), he still prioritizes statistical methods with respect to the goal of testing the hypothesized direct and indirect consequences of a postulated underlying process (*ibid.*: 12–3)—a view that he restated in his manifesto for sociology as population science (see Goldthorpe 2016: ch. 9), where, however, differently from early writings, he explicitly mentioned "agent-based computational modeling" as a possible methodological option for studying models of social mechanisms.

Interestingly, Ermakoff (2019) found, inductively, similar conceptual distinctions and methodological associations by reviewing scholarship in comparative and historical sociology. In particular, he observed that three different types of causal investigation are at work in this field: (i) a "morphological" approach, which bases causal claims on the detection of spatial and temporal patterns of social phenomena by relying on various descriptive techniques of data reduction (*ibid.*: 3–5); (ii) a "variable-centered" mode of causal analysis that infers causal claims "from patterns of association among a set of empirical categories" (*ibid.*: 6) by relying on multivariate statistical analyses that "probe the statistical significance of correlations and imputed effects" and comparative analyses detecting "combinations of attributes across cases"; and (iii) a "genetic" approach that "apprehends causality through the systematic investigation of generative processes" (*ibid.*: 11), a mode of investigation that Ermakoff connects to various methodological approaches to theorize and validate models of mechanisms, among which are "agent-based simulations" (*ibid.*: 16). Thus, on the one hand, what Ermakoff calls "morphological" and "variable-centered approach" squares with Goldthorpe's accounts of causality as "robust dependence" and "consequential manipulation"—Ermakoff explicitly evokes "the counterfactual framework" as a tool for "causal diagnoses" within the "variable-centered" approach (*ibid.*: 10)—and, on the other hand, Ermakoff's "genetic" category precisely correspond to Goldthorpe's type of causation as "generative process".

The association between specific methods for causal inference and views on causation—an association recently also documented by Kaidesoja (2021a)—is especially visible in the recent and flourishing literature on the so-called "potential outcome" approach. This is driven by the ambition to introduce the perspective of randomized experiments into the analysis of data generated outside an experimental setting (for a historical overview, see Imbens and Rubin 2015: ch. 2). Accordingly, the main task of the analysis becomes to show that individuals (or other units of analysis) that are exposed to different treatment states are likely to exhibit different responses, or outcomes. The causal effect of a given treatment state is then conceived as the (average) difference between the outcome of those who were exposed to it and that of those who were not.

In this way, the potential outcome approach is in essence tied to a counterfactual understanding of causation, which, as I have noted above, falls within the *dependence* (or *difference-making*) account of causation. From this perspective, establishing causal claims indeed amounts to quantify what-if outcomes, i.e. how a given group of units of analysis would have responded had their treatment value been different. As noted by Morgan and Winship (2015: 4), what-if (or potential) outcomes are counterfactual in the sense that they "exist in theory but are not observed". This is an important observation. It implies that, by construction, establishing causal claims within the counterfactual approach cannot be seen as a pure matter of empirical data. I will develop this line of reasoning further in Chapter 5 (see Section 5.3).

For the moment, the important point here is that the potential outcome approach, with its counterfactual understanding of causation, is now regarded by many as a "unified framework for the prosecution of causal questions" (Morgan and Winship 2015: 3). As such, it is seen as a tool that allows one to recast traditional multivariate statistical instruments in the terms of this particular view of causality. In this regard, Morgan and Winship's discussion of matching and regression estimators (2014: chs. 5–7) is especially illuminating. They elegantly show how the classic method of controlling for confounders can be reinterpreted as aiming not so much to identify "robust dependences"—to go back to Goldthorpe's distinctions—as to render comparable the outcomes of group subjects that were not randomly assigned to the treatment state of interest (see also Hernán and Robins 2020: ch. 15).

The diffusion of the potential outcome approach had an additional important consequence. It made explicit a conceptual distinction that helps to recognize that scholars
understanding causality in terms of dependence (rather than production) relationships
can themselves raise different causal inference questions. The distinction is that
between what Gelman (2011: 955) proposed to call "forward" and "reverse" causal
inference. According to him, when one is pursuing "forward" causal questions, one
seeks to understand and quantify "what may happen if we do X" whereas, when one is
interested in "reverse" causation, one wants to answer the question of "what causes Y".
In the former case, one focuses on one specific phenomenon (education, for instance)
and wants to establish the consequences of its presence, absence or variation (on fertility, for instance) whereas, in the latter case, one observes a given outcome and *a poste-
riori* wants to trace back the outcome to the various phenomena that may have made it
happen. For this reason, by using a terminology already present in John Stuart Mill's
(1882) *A System of Logic, Ratiocinative and Inductive* (see in particular chs. 6, 7, and 10),
the expressions "effects of causes" and "causes of effects" are also now often used to
refer to "forward" and "reverse" causation respectively—the latter in fact being often
also called "backward" causation (see, for instance, Sampson et al. 2013: 3, 7, 24). The
point I want to stress here is that, as Gelman (2011: 956) acknowledged, the potential
outcome approach is precisely the framework used by statisticians, and many economists, to treat "forward" or effect-of-a-cause questions, and experiments are seen as the
prototypical method to address this type of causal inference question (see also Dawid et
al. 2014). In contrast, many social scientists continue to address "reverse" or "cause-of-
effect" questions through multivariate statistical techniques, a choice that the potential
outcome perspective invites to question in terms of the assumptions that are needed to
establish the "causal" nature of each "cause" and its relative weight (see Gelman and
Imbens 2013). Interestingly, this contrast echoes an old intuition that Mill (1882: 557)
already expressed in the following way: "since, as a general rule, the effects of causes are
far more accessible to our study than the causes of effects, it is natural to think that this
method has a much better chance of proving successful than the former". The method
Mill was referring to precisely amounted to intervene on, or manipulate, experimentally
the factor of interest.[4]

Now, although it seems unquestionable that the potential outcome approach to causal
inference, and its associated effect-of-a-cause view, forces quantitative scholars to think
more rigorously, and more modestly, about how statistical methods for observational
data can be used to defend causal claims, it should be pointed out that the causality
account embedded within the potential account approach is still highly specific. In terms
of the aforementioned philosophical distinctions, the counterfactual view of causation
associated with the potential outcome approach is a type of *dependence*, or *difference-
making*, account of causality. From within a "production" perspective, this counterfactual view may be regarded as limited, in the sense that, in the words of the statistician

---

[4] I should thank Christopher Winship for pushing me to discuss explicitly the distinction between
"forward" and "reverse" causal inference, and articulate it more clearly with my double distinction
between dependence and production accounts of causality, on the one hand, and horizontal and vertical
mechanisms, on the other hand (see, on this point, my remarks in Section 1.4).

David Cox (1992: 297), it lacks "an explicit notion of an underlying process or understanding at an observational level that is deeper than that involved in the data under immediate analysis". To this, Cox adds: "my preference, however, is to restrict the term [causality] to situations where some explanation in terms of a not totally hypothetical underlying process or mechanism is available". Remarkably, Cox's critique is reminiscent of oldest statements by the realist philosopher Rom Harré (1972: 115–9, 136–7), who, already in the 1970s, remarked that the "successionist" view of causality should be complemented with a "generative" theory of causality. In Harré's (1972: 137) words, "science is based upon the generative theory, and treats the statistical evidence of succession as the basis for the hypothesis that a causal mechanism exists. This generates a methodological principle, in that a study is deemed complete only when the causal mechanism has been identified (...)".

Thus, similarly to philosophers of science, sociologists and statisticians clearly subscribe to different accounts of causality (for a clear synthesis, see Brady 2011: table 49.1). These different accounts tend to come with equally clear qualitative judgment on the respective merits of the various accounts. Goldthorpe (2001: 8–9), for instance, clearly states that the view of causation as "generative process" should be seen as an improvement on the "robust dependence" and "consequential manipulation" accounts because "it would appear to derive, rather, from an attempt to spell out what must be added to any statistical criteria before an argument for causation can convincingly be made". Hedström (2009), similarly, remarks that only the presence of a fully fledged mechanism authorizes causal inference and allows one to reach explanatory depth. To be sure, scholars within the potential outcome tradition would find this priority judgment unjustified because, so they would claim, mechanism-based explanations can be easily formulated within a counterfactual view and tested by an appropriate use of statistical methods (see Morgan and Winship 2015: ch. 10). As I will show next, however, different understandings of the concept of mechanism are at work here, thus increasing further the probability of misunderstanding and miscommunication between scholars that already understand causality in different manners.

1.3 *Horizontal* and *Vertical* Accounts of Mechanisms

As we have seen, "production" accounts of causality—differently from "dependence" ones—require the identification of an underlying mechanism for inferring causality from data. But what is a mechanism exactly? Similarly to the concept of causality, the concept of mechanism, too, has received a variety of interpretations (in philosophy, see Reiss 2013: 104–5; in sociology, see Mahoney 2001: 579–80; Hedström 2005: 25; Gross 2009: 360–2; in political science, see Gerring 2008). As recently observed by Kalter and Kroneberg (2014), the term mechanism has clearly penetrated much empirical research in sociology but it is still employed with a variety of meanings. Mapping this variety is important because, depending on how a mechanism is understood, ABM will be seen as either necessary to study models of mechanisms or unnecessary, and, consequently, different judgments will be formulated on the potential contribution of ABM to causal inference.

Again philosophical scholarship on mechanisms provides useful theoretical coordinates to appreciate various views on mechanisms (for an overview, see Andersen 2014a, b). For my purposes, the most relevant distinction here is between what I propose to call the "*horizontal*" and "*vertical*" views of mechanisms.

According to the former view, a mechanism is interpreted as a network of variables that stand in particularly robust relations. Woodward (2002: S375) exemplifies this view when he defines a model of a mechanism as a description of "(...) (i) an organized or structured set of parts or components, where (ii) the behavior of each component is described by a generalization that is invariant under interventions, and where (iii) the generalizations governing each component are also independently changeable" (for a discussion of the "invariance" and "modularity" conditions, see Kaidesoja 2021b). In contrast, according to the *vertical* view, a mechanism is envisaged as a "complex system" (Glennan 2002: S344) comprising a set of unities—entities and activities (Machamer et al. 2000), or component parts and operations (Bechtel and Abrahamsen 2005), or parts and interactions (Glennan 2002)—that, by interacting over time, generate some behavior of the system. Machamer et al. (2000: 3) exemplify this view when they define a mechanism as "(...) composed of both entities (with their properties) and activities. The organization of these entities and activities determines the ways in which they produce the phenomenon."

Although the vertical view is not incompatible with paying attention to the robustness of the relationships between the interacting parts that compose the mechanism (for an elegant discussion of this complex nuance, see Baumgartner et al. 2020), the distinctive feature of a mechanism from a vertical perspective is the dynamic of the changes a mechanism brings about. From their activity-centered perspective, Machamer et al. (2000: 3) put this point by saying that "entities often must be appropriately located, structured, and oriented, and the activities in which they engage must have a temporal order, rate, and duration" and that "description of a mechanism describes the relevant entities, properties, and activities that link them together, showing how the actions at one stage affect and effect those at successive stages" (*ibid.*: 12). From his interaction-centered perspective, Glennan (2002: S344) makes the same point when he remarks that "[a] mechanism operates by the interaction of parts. An interaction is an occasion on which a change in a property of one part brings about a change in a property of another part." In a word, what matters to the vertical view is the sequence of micro-level changes that dynamically create new connections within the system under scrutiny.

Thus, through the choice of the terms *vertical* and *horizontal*, I primarily intend to grasp the idea that, within the vertical view of a mechanism, the supposedly causal connection of interest is seen as gradually created by combining progressively the small changes triggered by the activities and the interactions of the low-level entities underlying the connection of interest. From this perspective, a model of a mechanism is conceived as a detailed account of *how* one moves from the low (small) level (scale), whatever it is, to the higher (or larger) level (scale). Andersen (2014a) expressed a very similar idea when she noted that treatments of mechanisms in the style of Glennan or Machmaer are intrinsically "hierarchical": the emphasis is on how nested entities and activities create connections between different levels of analysis. In contrast, within the horizontal view of a mechanism, the goal of providing a granular account of the changes that bridge levels of analysis (or different scales) is put into parenthesis (on granularity, see Chapter 2 and Chapter 4, Section 4.2.2). Andersen (2014b: 286) formulated this idea by noting that accounts of mechanisms *à la* Woodward are intrinsically "flat" in the sense that "the variables are all at similar levels (of size, organization, or other level differentiation)".[5]

With these qualifications in place, I will show next that sociologists, although without using the philosophical terminology, have engaged in vivid discussions about the merits and limitations of the horizontal and vertical accounts of mechanisms since the 1970s.[6]

1.3.1 Vertical *versus* Horizontal View

Among quantitatively oriented scholars, the confrontation between different approaches to mechanisms already appeared in the muscular critique Hauser (1976) addressed to Boudon's (1974) study of the temporal link between inequality of educational opportunities and social mobility in Western countries. Hauser's most general point was that Boudon did not make use of the best-developed framework for multivariate causal modeling at the time, namely path analysis. According to Hauser, the results that Boudon produced through numerical simulations actually depended on fragile and poorly validated assumptions. Although Boudon (1976) acknowledged that some of Hauser's methodological objections were appropriate, he essentially retorted to Hauser that he missed Boudon's main goal, which was to explore alternative research avenues "to go beyond the statistical relationships to explore the generative mechanisms responsible for them" (*ibid.*: 1187). Boudon's alternative consisted in "ideal-typical models" detailing how the aggregate patterns of interest—which were only summarized, but not explained, by statistical estimates, argued Boudon (1976: 1176, 1178–9, 1183)—can emerge from the dynamic and probably nonlinear relation among the actors' choices and their reactions to other actors' choices, as well as structural constraints (*ibid.*: 1180, 1185–6). A vertical

---

[5] It should be clarified that this does not mean that horizontal accounts of mechanisms do not, or cannot, consider different levels of analysis. Qualitatively similar information on a certain type of units of analysis can obviously be used to sort these units on various criteria so that clusters of entities are conceptually and numerically represented. In particular, through this procedure, different levels of analysis are represented under the form of different levels of aggregation. This is the typical *modus operandi* within multi-level models for nested data, for instance (see Gelman and Hill 2007). The point rather is the absence within a horizontal perspective on mechanisms of a specific and explicit substantive model in terms of activities and interactions explaining how transitions across levels operate.

[6] On the choice of the terminology "horizontal" and "vertical" accounts of mechanisms, let me add a further qualification. These two adjectives appear sometimes in the critical realism literature to identify a similar distinction as mine between an understanding of social causation based on associations between events and an understanding of causation where these associations are tied to activities and interactions of various entities at various levels of analysis among critical realists (see, for instance, Archer 1998: 196–7). However, this semantic similarity hides an important difference between my view and the critical realist perspective. To me, the notion of level is intrinsically analytical. That is why I constantly adopt the term "levels of analysis" or "levels of abstraction". Thus, the fact that I speak of a "vertical" view on mechanisms should not be taken as a sign of my commitment to a view of social reality as ontologically stratified, a view that I explicitly reject (on the difference between critical and analytical realism, see Di Iorio and Léon-Medina 2021).

view of mechanisms was clearly, but implicitly, at work here. Interestingly, as made more evident by a later article, Boudon (1979) regarded numerical simulations as a necessary tool for this alternative mechanism-based research strategy, although the type of simulations he employed were not, technically speaking, ABMs (on this point, see Manzo 2014b: 435–7).

Recent sociological scholarship shows that the bone of contention is still the opposition between the vertical and horizontal view of mechanisms that was behind the Hauser–Boudon debate. In one of the first meta-theoretical discussions on how the concept of mechanism may reorient empirical research in sociology, Pawson (1989: 130–1) noted that, although a mechanistic representation may have the cognitive function of making a connection between quantitative variables intelligible, it should not be conceptually equated with, nor methodologically operationalized as, a statistical control and/or a set of intervening variables.

This view animated the well-known volume on social mechanisms edited by Hedström and Swedberg (1998), which launched a new wave of discussions on mechanism-based explanations in sociology. As correctly noted by Mahoney (2001: 578, italics added), this new wave of mechanism-based thinking was explicitly motivated by the ambition to go beyond correlation analysis and by the rejection of the view that a mechanism can be simply understood "as an *intervening variable* or set of *intervening variables* that explain why a correlation exists between an independent and dependent variable". Hedström and Swedberg (1998: 17) indeed went back to the Hauser–Boudon debate, attacked the path-analytical tradition in sociology, and ultimately subscribed to the claim that "sociologists in the multivariate modeling tradition still make only rhetorical use of the language of mechanisms". Hedström's (2005: ch. 5) manifesto for analytical sociology endorsed a similar view. Hedström and Ylikoski's (2010: 51–2) review of causal mechanisms in the social sciences also acknowledged the variety of existing accounts of mechanisms but clearly expressed skepticism against the view that mechanisms consist of networks of intervening variables. This positioning is clear when Hedström and Ylikoski (2010: 54) explained that Woodward's (2002) counterfactual account of mechanisms is insufficient because, in fact, "[a] mechanism tells us why the counterfactual dependency [between cause and effect] holds and ties the relata of the counterfactual to the knowledge about entities and relations underlying it", a comment that nicely shows the opposition between Hedström and Ylikoski's endorsement of the vertical view in contrast to horizontal views in the spirt of Woodward.

Kalter and Kroneberg (2014: 101, italics added) echoed this opposition when they noted that, in much quantitative empirical research published in leading European sociological journals, "mechanisms as intervening variables" are "*mistakenly* seen to 'explain' the presumed causal effect of an independent variable on a dependent one". Kalter and Kroneberg lucidly recognized that "mediation analysis"—a crucial statistical tool among those who are animated by a horizontal understanding of mechanisms (for a recent brilliant presentation, see Makovi and Winship 2021)—can be used to test, at least partially, mechanism-based explanations but this requires, they argued, that the postulated mechanism has been previously designed theoretically in a clear and explicit way. In a word, Kalter and Kroneberg concluded, "Mechanisms should not be confused with potential indicators for potential concepts within potential mechanisms."

1.3.2 Horizontal *versus* Vertical View

Within the "horizontal" camp, too, skepticism was overtly expressed against this vertical account of mechanisms as hierarchical dynamic systems of entities, activities, and inter-actions. After all, as Morgan and Winship (2015: 330) remind us, "[f]or decades, social scientists have considered the explication of mechanisms through the introduction of intervening and mediating variables to be essential to sound explanatory practice in causal analysis". The counterfactual approach to causality—a typical case of the depend-ence account of causality, in the previous section's terminology—is now reshaping this methodological tradition. As a consequence, the concept of mechanism as a network of intervening variables is also being reshaped in these terms. Kaidesoja (2021b), for instance, explicitly distinguishes between a classical understanding of mechanisms as intervening variables, which he associates to the concept of causality as "robust dependence" (i.e. based on controlling for confounders), and a new version of this understanding of mecha-nisms as intervening variables where mechanisms are rather seen as sets of counterfac-tual dependences between variables, which Kaidesoja associated with Woodward's counterfactual interpretations of structural equation models.

This perspective change is especially visible in Knight and Winship (2013: 278). They regard the existing definitions of mechanisms from a vertical viewpoint as "unsatisfacto-rily vague" and propose a definition that, in their view, better clarifies in what sense a mechanism has a causal structure. According to this view, mechanisms are "modular sets of entities connected by relations of counterfactual dependence" (*ibid.*: 283). The modu-larity requirement, and the associated condition of "invariance" verified through "ideal" interventions (see Woodward 2003: 98), clearly refer to Woodward's (2002) manipulation-ist understanding of mechanisms that I discussed above as a typical case of horizontal accounts of mechanisms. In operational terms, however, Knight and Winship (2013: 282, emphasis added) continue to view a mechanism as "(...) a causal relationship involving one or more *intervening variables* between a treatment and an outcome", and they ulti-mately propose directed acyclic graphs as a framework for discussing under what condi-tions a net of "mechanistic variables"—the term appears in Morgan and Winship (2015: 335)—allows one to identify causal effects.

From the horizontal viewpoint, dissatisfaction with the vertical view of mechanisms is not only conceptual but also methodological. In this respect, two slightly different, although related, objections may be found in the literature. The first objection is that it is unclear what it means to empirically evaluate alternative hypotheses on mechanisms when mechanisms are regarded as dynamic complex systems of interacting entities bridg-ing levels of analysis. Morgan and Winship (2015: 345) are explicit on this point when they object that the mechanism movement—they refer here to Goldthorpe's (2001) and Hedström's (2005) proposals—runs the risk of falling prey to a "mechanism anarchy", i.e. a proliferation of mechanistic models, with no clear-cut proofs of their empirical signifi-cance, or, alternatively, a "mechanism warlordism", i.e. a proliferation of mechanistic models mainly supported by the scientific reputation of their proposers. As a remedy, they suggest a division of labor according to which the "generative mechanism movement", in their own words, contributes to causal inference by developing "how-possible" and "how-plausible" models, while "causal analysis", meaning quantitative techniques for observa-tional data from within a potential outcome approach, provides the tools for assessing the claims implied by models, which have the pretension to describe actual mechanisms (*ibid.*: 346–7). This proposal is further elaborated by Makovi and Winship (2021) who emphasize now the importance of mediation analysis as a tool to assess rigorously "what portion a variable's effect on an outcome is explained by one or more mediating variables", and reshape the meaning of a mechanism as the set of variables that "mediate some portion of the effect of T on Y". In this sense, they argue, a "mediating mechanism" (M), their own term, "unpacks the black-box of a treatment to outcome relationship by elaborating on how the causal effect is brought about (via M)".

These methodological proposals are in fact motivated by a second objection, which those who regard mechanisms as chains of intervening variables raise against those who regard them as dynamic complex systems of interacting entities bridging levels of analysis. This objection concerns the supposed lack of reliability of computer simulations as a method for building arguments for the existence of the postulated mechanisms (an objection that I will discuss carefully later, see Chapter 4, Section 4.3, and Chapter 5, Section 5.4). Morgan and Winship (2015: 341, fn. 15) formulated this skepticism overtly when they claimed that simulation seems at best a tool for "theory construction", and, even to this task, the tool is of limited utility because of its alleged lack of transparency. One may note that, although this criticism is not supported by a careful discussion of computational simulations, the mistrust of this approach within the "horizontal" camp is not new. By commenting on Hedström and Swedberg's (1998) early volume on social mechanisms, Morgan (2005: 26) already objected that "Sorensen and others got it only partly right. Without a doubt, they correctly identified a major problem with quantitatively oriented sociology. But, they did not offer a sufficiently complete remedy." In short, so the objection goes, no matter how appealing may seem the view of mechanisms as dynamic complex systems of interacting entities bridging levels of analysis, this is a good idea without a sound methodology.

1.4 Causality and Mechanism Accounts, and ABM's Perception

The analysis of the variety of accounts that the concepts of causality and mechanism have received clearly suggests that the different intuitions on the two notions connect in a systematic way. Scholars who have a dependence intuition about causality tend to see mechanisms as chains of intervening variables (horizontal view). In contrast, those who have a production intuition about causality tend to understand mechanisms as complex systems of interacting lower-level units that trigger higher-level outcomes (vertical view). This association is consequential because different views on causality and mechanisms tend to correspond in turn to different ways to open a "black box" underpinning a cause–effect connection, which ultimately lead to different appreciation of the potential contribution of ABMs for causal inference.

In particular, on the one hand, dependence accounts of causality and horizontal views on mechanisms consider that opening a black box amounts to uncovering intermediate variables between a treatment and an outcome. Within this perspective on causality and mechanisms, one should rely on quantitative tools that prioritize finding non-spurious relationships, establishing counterfactual claims, and, when possible, estimating unbiased parameters that quantify such relations, in a way that allows for the extrapolation from a sample, or test population, to an unobserved target population. As shown by the philosopher Peter Menzies (2012), the horizontal view is indeed typical of the literature on structural equation models and causal graphs (see Pearl 2009). From this perspective, ABM may seem unnecessary to establish causation: what matters is data quality and how creatively one is able to describe these data.

On the other hand, production accounts of causality and vertical views on mechanisms consider that opening a black box means to break the system down into parts and show that the dynamic of the interactions between them can generate, in the sense of reconstruct, the aggregate behavior under scrutiny. Within this perspective, one should rely on methods that prioritize finding a credible narrative that accounts for the observed patterns. The idea is that dependence relations are not constitutive of causality but rather the manifestation of it. From this point of view, simulation methods, and ABMs in particular, would thus seem powerful tools for studying the details of these complex dynamics behind observed connections between happenings. ABM appears as a crucial tool to establish causation (in the sense of production accounts of causality): it provides a formal device to prove that the dependence relationship under scrutiny is deducible by unfolding the postulated (formalized) narrative (see Anzola 2020: 55).

The different reactions to Lucas's (1976) critique to causal inference in macroeconomics (concerning the claim that inflation causes employment) provide a nice historical illustration of the difference between the two camps. In the "dependence" camp, there was a data-driven reaction, which emphasized the centrality of intervention-like methods and led to a more sophisticated use of statistics, the diffusion of time-series econometric models, and the development of vector autoregression methods (Sims 1980), in the tradition of Granger (1969). In contrast, in the "production" camp, there was a theory-driven reaction, which demanded that macroeconomic models be enriched with "micro-foundations". This led initially to the intense use of rational choice models calculating economic aggregates based on individual preferences and expectations—a development encouraged by Lucas (1976) himself—and, consequently, to the critique of representative–agent assumptions (see Kirman 1992; Hoover 2008a, b), to introducing agent-based computational models for solving the aggregation problem in the presence of actors' heterogeneity and various types of social networks (Tesfatsion 2002, 2006; Arthur 2006, 2021). Sociology, and analytical sociology in particular, followed the same path by moving from regression-like statistical methods for survey data to agent-based computational models in order to address the micro-to-macro problem when interdependence structures are present (see Manzo 2020: 200–4).

Some readers may find this opposition exaggerated. Certain ways of framing relationships between different types of causal reasoning may indeed give the impression that the vertical and the horizontal views of mechanisms in fact are closer than it may seem at first glance. This can be seen by looking into the way some scholars have proposed to combine "forward" and "backward" (or "reverse") causal inference (on this distinction, see Section 1.2).

In particular, Sampson et al. (2013) argued that, once the effect of a given cause has been established, we are left with the question of "why" the treatment generated the observed outcome. Thus we need what they call "causal interpretation". To perform this task, Sampson and colleagues suggest, we have to go back to "backward causation" and design the possible set of causes to which the effect of the specific cause that we have documented belongs. Sampson et al. (2013: 7–8) go until using the term "mechanistic causality" (as a complement of counterfactual causation) to refer to this operation of "causal interpretation" where one tries to unpack the black box underlying the "first-order" effect-of-a-cause dependence documented through a manipulationist approach. This is especially important, Sampson et al. argue, when the goal is to translate the effect of a treatment into policy interventions because, for this goal, we must know why the treatment produces the effect, how this dependence depends on context, and how the effect of the treatment varies across subgroups and over time. Gelman and Imbens (2013) developed a similar line of reasoning when they suggested that "why" questions associated with a causes-of-an-effect approach should be combined with "what-if" questions associated with an effect-of-a-cause perspective. Asking "why" questions, they argue, allows explanatory hypotheses to be formulated for understanding a given documented effect-of-a-cause link as well as to discover errors in the model specification adopted to identify the supposed causal link.

This emphasis on "why" questions may thus seem equivalent to the "why" questions that motivated the "new" vertical view of mechanisms in the philosophy of science and in analytical sociology, which also wants to understand "how" a given connection of interest is brought about. However, the qualitative similarity in terms of cognitive goals— in the sense that, in both cases, one is driven by a need for psychological understanding of the observed connection—translates in different methodological solutions about how the "causal interpretation" should be operated. All examples given by Sampson et al. (2013: figs 2–10) actually amount to "disaggregating" (their own word) the effect-of-a-cause into a causal graph of potentially mediating variables; similarly, Gelman and Imbens (2013) proposed to study the potentially explanatory net of causes-of-the-effect through the same potential outcome approach that was adopted to document the first-order effect-of-a-cause dependence. The possibility of building, and simulating, a formal model of a dynamic system composed of activities, entities, and interactions that could be exploited to show the conditions under which the causal link of interest can appear is not even mentioned.

As the reader knows, no matter how clashing these views on causality, mechanisms, and legitimate methods for causal inference may appear at first, this essay is motivated by the conviction that these views in fact can, and should, be reconciled. Later on, especially in Chapters 4 and 5, I will accumulate elements that suggest that proper causal inference requires a combination of dependence (horizontal) and production (vertical) accounts of causation (mechanisms), thus a synergy between experimental and statistical methods for observational data on the one hand and ABM on the other hand. I will fully develop this argument when defending the "evidential pluralism" thesis (see Chapter 6).

For now, however, let us take at face value the contrasts observed in the literature on causality and mechanisms, and try to explain why, among mathematical models and simulation methods, ABMs can be seen as having a special value for modeling mechanisms from a production point of view on causality and a vertical perspective on mechanism.

2

Agent-based Models and the Vertical View on Mechanism

Are there specific technical arguments supporting the claim that an agent-based model (ABM) is particularly flexible for designing models of dynamic systems of interacting entities bridging levels of analysis, thus justifying the view—implicitly assumed in the previous chapter—that ABMs are especially compatible with a production account of causality and a vertical view of mechanisms?

From a methodological perspective, one may in fact object that some statistical techniques, such as multilevel regression (Gelman and Hill 2007) or stochastic actor-oriented models for network dynamics (Snijders and Steglich 2015), or mathematical formalisms such as recursive Bayesian networks (Casini et al. 2011; Clarke et al. 2014) and game-theoretic models (Raub et al. 2011) also presuppose a vertical understanding of mechanisms in that they explicitly deal with the connection between micro- and macro-levels of analysis. On the other hand, other computational techniques, too, for instance cellular automata, artificial neural networks, or genetic algorithms, are "bottom-up" and focus on the behavior of single entities (Gilbert and Troitzsch 2005: chs. 7, 10); as such, they, too, may be used to operationalize the vertical view of mechanisms.

Thus what is special, if anything, about ABMs? With respect to statistical methods, the general point is that an ABM can recreate dynamically the sequence of the events responsible for the connections between the levels of analysis of interest rather than summarizing these events through a set of estimated parameters (on this point, see Stadtfeld 2018). With respect to other mathematical or computational techniques, some studies in computational biology (Zhang et al. 2009; Wang et al. 2013) and economics (Hayward 2006) have suggested that an ABM can incorporate all of the modeling advantages of such techniques and thus provide a more general and powerful tool (for a review of coupling heterogeneous models within a single ABM, see Morvan 2013).

More specifically, it can be argued that an ABM allows a greater amount of flexibility, granularity, and generality for the implementation of the vertical view of mechanisms than statistical and other computational methods. By flexibility, I mean that an ABM is not restricted to model any specific kind of entities, properties, activities, interdependence structures, levels of analysis, sequence of activation, or behavioral rule (Axtell 2000). By granularity, I mean that an ABM does not restrict a priori the level of detail at which one can describe each of these elements. By generality, I mean that an ABM can include several formalisms, each of which can be used to model a

Agent-based Models and Causal Inference, First Edition. Gianluca Manzo.
© 2022 John Wiley & Sons, Inc. Published 2022 by John Wiley & Sons, Inc.

specific aspect of the mechanisms under scrutiny—this feature of an ABM has been called "pluriformalization" (see Varenne 2009: 14).

In this chapter, I briefly explain why an ABM's deep technical infrastructure in fact is responsible for an ABM's high flexibility, granularity, and generality, and thus makes it a powerful tool for the implementation of a vertical view of mechanisms. This chapter focuses on an ABM's intrinsic technical potentialities. The practical difficulties one must handle for realizing such potentialities will be discussed later (see Chapter 4).[1]

2.1 ABMs and Object-oriented Programming

In essence, an ABM is a computer program designed to formally represent a set of hypotheses and deduce, in a numerical form, the logical implications of such hypotheses. Since the program is written from scratch, there is no *a priori* constraint on the substantive content of the hypotheses the modeler wants to implement. This flexibility is especially reinforced by the particular type of programming language that can be used to design an ABM.

The smallest units of an ABM are "objects", which from a computer science viewpoint are "computational entities that encapsulate some state, are able to perform actions, or methods, on this state, and communicate by message passing" (Wooldridge 2009: 28). That is why de Marchi and Page (2014: 1) define ABMs as consisting of "autonomous, interacting computational objects, called agents, often situated in space and time" and Macy and Flache (2009: 248) note that an ABM "replaces a single integrated model of the population with a population of models, each corresponding to an autonomous decision maker". Any single object can indeed be seen as a computer program (see Wooldridge 2009: 5).

The deep connection between object-oriented languages and ABMs lead some authors to use the label "agent-based object modeling" (Miller and Page 2007: 78) or "object-oriented simulation methodology" (see Hummon and Fararo 1995a: 8; see also 1995b) instead of that of ABM *tout court*. To be sure, in theory, any programming language containing minimal requirements can be used to program an ABM (on this point, see Izquierdo et al. 2009; Nikolai and Madey 2009). However, it is widely agreed that, in practice, complex ABMs are difficult to build without using object-oriented programming, which proves the existence of an intimate link between this programming style and an ABM's flexibility (Shalizi 2006: §5).

Within object-oriented programming, the modeling process indeed amounts to reformulating the explanatory hypotheses (i.e. the *explanans*) into a set of "classes" of objects, namely groups of objects that share the same properties and functions, and arranging them in such a way that the behavior of objects in one class constitutes the input for the behavior of objects in another class. Studying the *explanans* amounts to simulating its computational implementation, which practically means updating the attributes attached to the objects that make up the ABM, iterating the rules that define the objects, and letting the objects communicate—hence influencing each other—over the simulated time. Since

---

[1] Sections 2.1–2.6 of this chapter build on and extend Casini and Manzo (2016: 14–18).

objects are conceptually empty, any type of problem can be "objectified" at any level of required detail (granularity); and, since classes of objects can be given any specific type of behavior, models can be composed of sub-models driven by different types of modeling choices, thus making it possible to assemble different modeling options within a single ABM (generality).

When an ABM is viewed in terms of its fundamental computational components, namely objects, the affinity between the ABM and the vertical view of mechanisms becomes clear. Similarly to real-world social (or biological) mechanisms, which are constituted of entities (at several levels of organization) with their properties and activities, and are put in motion when these entities act and influence each other, ABMs are constituted of objects with their attributes and procedures (or methods, or functions) and are turned into dynamic processes when the objects are invoked and asked to execute the procedures attached to them. In sum, an ABM inherits from its object-oriented basis an internal structure that is homologous to the structure and the functioning of what one wants to study within a vertical view of mechanisms. By using unified modeling language (UML) diagrams, Vu et al. (2020) has recently provided a particularly clear graphical illustration of this intimate connection between ABMs, objected-oriented programming, and "vertical" mechanisms.

For the design of models of social mechanisms, ABMs' greater flexibility, granularity, and generality compared to statistical methods or other simulation-based modeling strategies should be especially appreciated with respect to five elements that have continuously challenged quantitative methods in sociology. These elements are heterogeneity, micro-foundations, interdependence, time, and multi-level settings. I now discuss each of these elements and explain how an ABM's capacity to model them relates to an ABM's object-oriented infrastructure.

2.2 ABMs and Heterogeneity

The design and manipulation of single computational entities, namely objects, implies that unrealistic modeling shortcuts such as averaging under the form of representative agents, so common in game-theoretical models for instance, are never required within ABMs (Gallegati and Kirman 1999). Objects can be heterogeneous in a variety of ways (Epstein 2006: xvi, 7). First, objects within the same class, whilst by definition sharing the same properties (and activities), are such that these properties can get different values. Second, objects in different classes by construction possess different types of behavior. Third, by playing with the objects' scheduling, objects can be represented as being heterogeneous in terms of behavioral sequences, i.e. the time at which a given behavior is realized. Finally, objects are conceptually "empty", meaning that by playing with variables, vectors, or other data structures the objects' states can model any attributes of the entities of interest, and by creating logical and numerical functions over these states the objects' methods (or functions) can be used to model every activity of these entities. In consequence, heterogeneity can take the form of multiple classes of objects representing different types of entities at different levels of abstraction, such as states, organizations, firms, or actors (Farmer and Foley 2009). As stressed by Miller and Page (2007: 84–5), homogeneity may be a convenient and theoretically legitimate assumption. The point is that an ABM does not constrain us *a priori* to impose homogeneity because of tractability issues that are unrelated to substantive considerations. Within ABMs, the amount, and the type, of heterogeneity to be assumed becomes a modeling problem itself, and the implications at a higher level of analysis of the choices made at a given level of analysis in terms of heterogeneity can be deductively studied in a systematic way.

2.3 ABMs and Micro-foundations

Whatever type of entities one considers at the micro-level of a given ABM, the method is highly flexible with respect to the type of behavior one wants to design. This arises again from the object-oriented nature of the technique. Objects are defined by the attributes and functions one attaches to them. Similarly to attributes, the objects' functions can also be of all sorts. Since the model is solved by simulation, there is no *a priori* constraint on the type (logical or numerical) and form (continuous or discrete) these functions can take. This allows a great deal of flexibility and granularity in designing the entities' behaviors. When objects are used to model individuals, a large spectrum of options is available to represent the actors' reasoning and choices, for instance simple heuristics (Miller and Page 2004: 10; Todd et al. 2005), heuristic-based game-theoretic strategies (Alexander 2007: 38–42; Gintis 2009: 72–3), sophisticated maximizing behaviors (Shoham and Leyton-Brown 2009), complex cognitive reasoning (Wooldridge 2000), or argument-based decisions (Gabbriellini and Torroni 2014).

Thus, contrary to the association that some—usually unfamiliar with the field—tend to establish between ABMs and rational-choice theorizing (see, for instance, Elster 2009a: §2), the tool is entirely agnostic on the kind of micro-foundations a modeler should subscribe to. In fact, ABMs can accommodate a large variety of cognitive mechanisms driving the actors' behavior (Miller and Page 2007: 81–3). Systematic review of agents' internal architectures also documented the impressive variety of options ABMs offer for modeling human decision-making (see Balke and Gilbert 2014). Moreover, insofar as many ABMs have shown that simple, and cognitively plausible, rules are enough to derive stable macro-equilibria on realistic time scales (Epstein 2006: ch. 1; Manzo and Baldassarri 2015), ABMs can even be used as formally supporting a critique of the lack of realism of models postulating sophisticated rational calculations at the micro-level (for a recent statement in this sense, see Arthur 2021).

2.4 ABMs and Interdependence

As we have seen, one of the central features of a vertical understanding of the concept of mechanism concerns the interactions between the elementary units of a system, and how these interactions gradually generate sequences of events and reactions that lead to the construction of the dependence relationship of interest. The object-oriented nature of ABMs again is behind ABMs' capacity to handle several forms of interdependence with great flexibility and granularity.

According to Wooldridge's aforementioned definition of objects, one of the features of computational objects is that they can communicate with each other (through message passing at the memory-address level). This is the programming aspect that helps us see why an ABM is so flexible in embedding entities' behaviors within more or less large interdependence structures (Epstein 2006: 6, 52). Since the attributes' values can travel from one object to another, it is easy to make one object's behavior depend on another object's behavior. This dependence can assume three general forms.

First, object interdependence can be mediated by a global aggregate, namely an outcome derived from the behavior of all objects present in the artificial population that feeds back onto the subsequent behaviors of each object. Second, object interdependence can be mediated by a local aggregate, namely an outcome derived from the behavior of all objects to which the focal object is connected that feeds back onto the subsequent behaviors of the focal object. Third, object interdependence can be purely dyadic, i.e. the relevant input for a given object comes from a single other object. In the last two cases, the relevant object's neighborhood can be defined on a spatial and/or relational basis. By exploiting the exchange of information at the deep level of the computer's memory addresses (Hummon and Fararo 1995a), ABMs thus allow great flexibility in designing space-based dependences (Miller and Page 2004) and/or network dependences (Rolfe 2014).

The important point here for sociological theorizing is that, by playing with the objects' attributes, functions, and communication, an ABM allows one not only to design spatial and relational structures but also to design mechanisms that clarify how such structures shape lower-level entities' components such as beliefs, opportunities, perceptions, or desires (see DellaPosta et al. 2015; Eberlen et al. 2017; Muthukrishna and Schaller 2020).

2.5   ABMs and Time

It has often been argued that sociological theorizing and quantitative empirical research have systematically under-elaborated the way time shapes social actions and interactions (see, respectively, Abbott 2001: ch. 7; Winship 2009). An ABM offers great modeling opportunities in this respect. Since an ABM is a simulation-based method, it is intrinsically dynamic (Miller and Page 2007: 80–1, 83–4). When the first set of objects is invoked to execute the procedures defining their behavior, a chain of activities, reactions, and updating is triggered, such that the final higher-level outcome is generated step-by-step by a concatenated cascade of local upward aggregations and downward effects. Thus, when an ABM is simulated, the mechanisms defining it are transformed into the unknown process potentially contained in these mechanisms. The point that needs to be stressed here is that time itself can be modeled within an ABM. As shown by Axtell's (2000) seminal article (but see also Miller and Page 2004), the order in which objects are invoked and updated, as well as the order in which procedures are executed by a given object, are important modeling choices. And, once again, an ABM allows the greatest flexibility in this respect (for a recent review of agents' scheduling options within ABMs of opinion dynamics, see Weimer et al. 2019). In this sense, an ABM provides sociologists with the possibility of exploring systematically, and thus reflecting upon, the higher-level consequences of different hypotheses concerning the temporal organization of actions and interactions.

2.6 ABMs and Multi-level Settings

As I explained in the previous chapter, the term *vertical*, as opposed to *horizontal*, was chosen to suggest that the *vertical* view of mechanisms has the ambition to demonstrate how a dependence relationship of interest is gradually created (or reconstructed) through progressive combinations of small changes across several levels of analysis, a feature that squares with a *production* view on causation. An ABM is particularly powerful in integrating different levels of analysis. This is true from both a static and dynamic perspective.

From a static point of view, as testified by the so-called "agent/role/group" architectures in computer science (Ferber et al. 2005), work on cancer growth in biology (Zhang et al. 2009; Wang et al. 2013), and research in computational organization theory (Carley 2002; Harrington and Chang 2005; Fioretti 2013), the conceptual emptiness of computational objects implies that entities as diverse as particles, molecules, cells, beliefs, actors, groups, organizations, and states can be modeled and coexist within a single ABM, thereby allowing the co-habitation of several levels of analysis.

From a dynamic point of view, by exploiting the objects' communication, the simulation of an ABM allows one to establish dynamic relations between these levels. In this regard, it is important to appreciate that, within an ABM, three different types of relations can be established across levels of analysis.

First, it is possible to create "lateral" connections. By this I mean interdependences among objects representing entities at the same level of analysis independently from the activities of objects representing entities at lower levels of abstraction. Second, "downward" relations are also possible, i.e. relations that establish interdependences between the behaviors of objects at a given level of analysis and those of objects at a lower level of analysis, or relationships involving local/global aggregates generated at time t that feed back onto the objects' behavior at time $t + 1$ (we saw this case while discussing interaction structures). Third, "upward" relations can be triggered, i.e. interdependences between the behaviors of objects at a given level of analysis and objects that collect the behavior of lower-level objects to compute the resulting outcome at a higher level of abstraction.

From the point of view of the history of quantitative methods, an ABM's capacity to implement "upward" interdependences among objects is especially important. As already remarked by Coleman (1986: 1316), statistical techniques for observational data had traditionally been strong at assessing the effect of group- and individual-level factors on individual-level outcomes (and, today, we may add the effect of network- and individual-level features on network-level outcomes), but they had not been equally efficient as "methods for characterizing systemic action resulting from the interdependent actions of members of the system". In contrast, when one simulates an ABM, the macro-level can be reconstructed step-by-step by iterating the objects' behavior, by making the objects communicate, and by collecting the local products of these behaviors over time. In this sense, Epstein (2006: 21) claims that "agent-based models allow us to study the micro-to-macro mapping", which clearly is a desirable feature for those who are animated by a production view on causation and a vertical perspective on mechanisms.

However, the methodological novelty of ABMs with respect to "upward" interdependences, and consequently the capacity of modeling the micro-to-macro transition, should not lead us to forget that an ABM is equally flexible about the implementation of "lateral" and "downward" connections among objects. This is a very important technical fact that must be recognized in order to appreciate that an ABM is not limited to its bottom-up "generative" capacity (for the temptation of equating ABMs to "generative fundamentalism", see for instance Little 2016: ch. 5). Given the various set of relationships across levels of analysis, and the variety of entities that an ABM can represent through computational objects, this method allows us to model a larger set of phenomena involving multi-level settings than what may be thought if an ABM was seen only through its bottom-up "generative" capacity (for a theoretical discussion of this point, see also Manzo 2020; for a new technical proposal for handling complex multi-level ABMs, see Hjorth et al. 2020).[2]

2.7 Variables within Statistical Methods and ABMs

Admittedly it is not immediate to grasp the object-oriented nature of ABMs and the deep implications of this programming paradigm for ABMs' flexibility, granularity, and generality. First-hand experience with object-oriented languages can probably help to see this connection. Among statistically minded social scientists, those who exploited object-oriented programming to create new statistical methods and/or software packages also tend to recognize an ABM's interest for causal inference (see Stadtfeld 2018; Amati and Stadtfeld 2021; Steglich and Snijders 2022). Thus one may suspect that limited first-hand familiarity with object-oriented programming partly explains the widespread skepticism among many statistically minded scholars about the claim that an ABM is capable of representing a mechanism in a qualitatively different way in comparison with a statistical model.

Doubts on an ABM's distinctiveness as a modeling tool for social mechanisms is indeed clearly visible among scholars endorsing a dependence account of causation and a horizontal view of mechanisms. According to Morgan (2005: 31), for instance, "[t]he appeal for mechanisms is a useful rallying cry, but the originality of a mechanism-based sociology has been oversold. (...) Arguing that mechanisms are concatenations of nonlinear functions is not an argument against the use of variables, since the primitive elements of functions—defined as inputs and outputs—can be redefined as variables." This statement deserves special attention because it could be used to argue that, since a theoretical representation of a mechanism requires variables and functions (something I entirely agree with), a structured set of intervening/mediating variables (i.e. a mechanism in the horizontal sense) can be considered a "mechanism sketch" (Morgan and Winship 2015: 346–52) and multivariate statistics a tool for directly testing mechanism-based explanations (on this point, see also Opp 2007: 121).

From the vertical view of mechanisms, however, this implication would be incorrect because it fails to appreciate that the variables, and the functions operating on them, play a different role within a statistical model compared to an ABM. In the former case, variables and functions are used to detect a pattern of average effects which may reflect the

---

aggregate statistical signature of the postulated underlying mechanism (made of entities, properties, and interactions). In the latter case, in contrast, variables and functions are used to represent precisely the properties, the activities and the interactions of the low-level entities of interest, and functions are additionally exploited to put in motion (i.e. to simulate) the dynamic associated with the postulated mechanisms with the goal of assessing under which conditions this dynamic brings about the aggregate dependence relationship of interest.

Thus, although it is indisputable that data structures (variables being one of them) and operations on these data structures (i.e. functions) are needed to operationalize a theoretical representation of a (set of) mechanism(s) within an ABM, it should also be clear that variables and functions are, within an ABM, the means to recreate *in silico* a series of small chains of events that gradually lead to a given connection of interest. In this sense, the fact that mechanisms (vertical sense) are built on, or through, chains of variables (i.e. mechanism in the horizontal sense) does not mean that the former are qualitatively reducible to the latter.

Scholars endorsing a dependence view of causation and a horizontal account of mechanisms may concede this point but, I suspect, they would formulate the following counter-objection: variables, within statistical models, are empirical constructs whereas, within ABMs, they are only logical and/or numerical features. As a consequence, although statistically minded scholars may be disposed to acknowledge that variables are used in two different ways within ABMs and statistical models, they would maintain that, for proper causal inference, even accepting a production perspective on causality, ABMs would be ultimately useless because they cannot tell anything about reality.

In the statistical literature, traces of this argument can be found, for instance, in Gelman and O'Rourke (2013: 1–2) when, after considering several sources of evidence (including "mathematical theory" and "computer simulations") that could be exploited to defend the effectiveness of a new statistical method, they claimed that "(...) theory and simulations are only as good as their assumptions (...)". Among sociologists, Goldthorpe (2016: ch. 7, fn. 7) commented more explicitly on ABMs and remarked: "This strategy can provide a strong test of the causal adequacy—or of what ABC modellers refer to as the 'generative sufficiency'—of a proposed mechanism (...). However, to repeat the point made in the text, to show the generative sufficiency of a mechanism is not to show that it is in fact this mechanism that is in some particular instance at work."

As I will demonstrate in the next chapter, this objection tends to ignore that ABMs in fact now come under very different forms. In particular, when an ABM is based on empirical calibration, many of the "variables" (and functions) on which the simulation is based are as "real" as those supporting statistical (or experimental) modeling, a fact that suggests that more refined arguments are required to disqualify ABMs for causal inference.

3

The Diversity of Agent-based Models

In the previous chapter, I discussed agent-based models (ABMs) in general by focusing on the technical reasons that make this computational method especially flexible for the design of mechanisms seen as multi-level dynamic systems of interacting entities (i.e. the vertical view of mechanisms, in Chapter 1's terminology). In this chapter, I drop this simplification and document the diversity of ABMs that exist in the literature. In particular, I focus on three structural dimensions that concern how the model is built and analyzed. These dimensions, I argue, all impact on the subjective confidence we may have in the reliability of the results produced by the model, and, consequently, on its potential contribution to the defense of a specific causal claim. These three dimensions are (i) the extent to which the assumptions that are used to design the model are explicitly grounded on (empirically supported) sociological and/or psychological theories (I call this dimension "theoretical realism"); (ii) the extent to which the assumptions that are used to design the model are calibrated by using empirical and/or experimental data (I call this dimension "input realism"); and (iii) the extent to which the model's outputs are confronted with empirical and/or experimental data (I call this dimension "output realism").

Although each of these dimensions should be seen as operating on a *continuum*, and the judgment on where a specific model locates on this continuum is obviously partly observer-dependent, an explicit discussion of the variability of ABMs along these lines is important because the large variety of ABMs on offer constitutes a third element—in addition to the diversity of views on the concepts of causality and mechanisms (see Chapter 1)—that contributes to explain the diversity of judgments on the role of ABMs for causal inference. Indeed, depending on the specific type of ABM considered, one can reach different conclusions on the extent to which this method can contribute to build persuasive causal claims.

Thus, in this chapter, I first reconstruct the historical roots of the variety of ABMs with respect to the links the modeler establishes between the model's assumptions, existing theories and real-world data (Section 3.1). By doing so, I show that, from the very beginning, the method has been equally employed in an abstract as well as in a data-oriented way, which suggests that ABMs are not intrinsically limited to be used as pure formalized abstractions. Then, by reviewing both pioneering works and more recent applications, I document a slow trend in the literature towards more realistic ABMs (Section 3.2). Finally, on the basis of a meta-analysis of a selection of models from several disciplines, I develop

Agent-based Models and Causal Inference, First Edition. Gianluca Manzo.
© 2022 John Wiley & Sons, Inc. Published 2022 by John Wiley & Sons, Inc.

a typology of ABMs that clarifies that, within the general trend of increasing ABM realism, "theoretical", "input", and "outcome" realism still often vary independently within a given ABM (Sections 3.3 and 3.4).[1]

This is an important observation that suggests that, since these three dimensions arguably all have an impact on the reliability of what we can learn from the ABM at hand on the targeted mechanism, a principled assessment of the usefulness of ABMs in causal reasoning cannot neglect the variety of ABMs available on the market. That is why, to anticipate what the course of my argument will be, in Chapter 4, rather than studying this or that ABM to defend this or that view on the method's potential contribution to causal inference, I will build on the dimensions discussed in this chapter to provide a general discussion of the sense in which, and the conditions under which, if any, ABMs can produce results that are relevant for defending persuasive causal claims.

3.1 Abstract *versus* Data-driven ABMs: An Old Opposition

When studies pioneering the ABM methodology in the 1960s are considered, it appears that the ABM has been used in different ways since the beginning of its history in the social sciences. To appreciate this point, let us go back to Thomas Schelling's (1971) acclaimed model of ethnic segregation and to a far less-known research by the Swedish geographer Torsten Hägerstrand (1965) on the diffusion of innovations (on the trajectory of Schelling's case, see Hegselmann 2017).

Schelling postulated an ideal uni-dimensional or—in the most famous model's variant— bi-dimensional space in which "stars" and "zeros" asynchronously decide to change their location as a function of their closest neighbors' features. Schelling's goal was to see whether, starting from a random distribution of stars' and zeros' locations, mild homophilous preferences as to the composition of one's close neighbors were sufficient to generate clusters of similar entities when relocation choices were repeated over time as a function of those preferences. To answer this question, Schelling varied several aspects of the model, like the intensity of preferences for like-neighbors and the size of groups and neighborhoods, and studied how spatial patterns changed as a function of these modifications (a procedure that is now known as sensitivity analysis, see Chapter 4, Section 4.3.1).

For our discussion, what matters is that Schelling did not use on the input side any specific sociological or psychological theory to justify theoretically his micro-level assumptions (i.e. more or less weak preferences for similar others); nor did he use any empirical data to set the preferences assigned to his abstract entities, or the entity's relevant neighbors, or still the size of the two groups. Instead, Schelling drew on a (weak) structural analogy, based on intuitions and common sense, between his fictional mechanism and the mechanism for segregation in reality. On the output side, simulated patterns were not confronted with empirical data on ethnic segregation in a specific geographical area; as Schelling himself admitted, his analysis was in fact intended to cover any phenomena in which two groups have some tendency to stay apart from each other (*ibid.*: 144, 158).

---

[1] This part of the analysis builds on and extends Casini and Manzo (2016: 19–28).

In a largely neglected, but this notwithstanding remarkable, study, the Swedish geographer Torsten Hägerstrand (1965) followed a different strategy. His *ante litteram* ABM was designed to account for patterns of temporal and spatial concentration of farm innovations in two Swedish regions. Hägerstrand's hypothesis was that adopters can contaminate potential adopters as an inverse function of the physical distance between them. To study this hypothesis, he designed an ideal bi-dimensional space in which "robots" (in his own words) meet and spread information as the inverse to the squared physical distance separating them. For my purpose, the crucial point here is that Hägerstrand distributed "robots" on the grid in a way that reproduced the distribution of farms in the Swedish regions of interest and set the matrix of dyadic contact probabilities on the basis of independent statistical sources describing local phone traffic and migration flows in the regions of interest. To assess the realism of the outputs generated by the simulation of the model, he then systematically compared the simulated temporal and spatial patterns of adoption with the Swedish actual data.

Thus, both Schelling's and Hägerstrand's agent-based simulations were based on a simple micro-level mechanism, namely weak distaste for dissimilar others and spatially constrained information transmission respectively. In both cases, the mechanism was intuitively believed to be general and realistic. Ultimately, this belief relied on common sense, and not on specific sociological or psychological theories on which some consensus existed in the scientific community at the time the model was formulated. In the terminology that I will introduce later, both models were weak in terms of "theoretical realism" (see Section 3.3 below).

Given this similar starting point, however, Schelling and Hägerstrand operated differently with the posited micro-level mechanism. Schelling's goal was to explore the space of its logical implications at the population level. The simulation was used to discover counter-intuitive consequences. The model's parameter and structure were partly modified to assess the robustness of the surprising outcomes discovered. After all, the realism of the model was of secondary importance. It was the model's heuristic value that really mattered to him. Hägerstrand's goal was different. He wanted to reproduce a specific portion of reality at a specific time and place. To increase the confidence in the posited micro-level mechanism, he anchored it to specific empirical values and used it to generate simulated patterns under realistic input constraints. Hägerstrand was not interested in exploring all the space of the possible logical macroscopic implications of the model. Instead, he aimed at incrementally refining the model until the simulated spatial patterns of adoption acceptably matched the actual Swedish data.

By operating in this way, Schelling and Hägerstrand implicitly outlined two methodological orientations that still deeply inform contemporary studies using ABMs. In their purest form, these orientations are illustrated by two principles that it is now common to call "Keep It Simple, Stupid" (KISS) and "Keep It Descriptive, Stupid" (KIDS), respectively.

Supporters of the KISS approach share Axelrod's (1997: 5) conviction that "if the goal [of ABMs] is to enrich our understanding of fundamental processes that may appear in a variety of applications (...), then simplicity of the assumptions is important, and realistic representation of all the details of a particular setting is not". Within this perspective, the target of the model typically is highly abstract: it usually consists of qualitative properties shared by a large set of phenomena. This form of to-be-explained outcomes corresponds to what Boero and Squazzoni (2005: §3.2) called "typifications". The term "stylized facts" (whose historical origin in economics is unrelated to ABMs) is also frequently used to qualify the target an ABM built within the KISS perspective (for a recent discussion, see Cramer and Trimborn 2019). KISS-oriented ABMs are seen as "tools to think with" (O'Sullivan and Perry 2013: 14–15) or "intuition engines" (de Marchi and Page 2014).

On the other hand, supporters of the KIDS principle believe that the ABM flexibility for mechanism design is so high that simple models should only be the possible result of the analysis, not its starting point (for a detailed discussion, see Edmonds and Moss 2005). More particularly, within a KIDS perspective, one should start with a model that is as descriptively rich as the available empirical evidence allows, and then simplifications can be introduced as long as a given simplification is not inconsistent with what is known about the empirical functioning of the phenomenon under scrutiny and as long as the simplification does not reduce the model's performance in reproducing the to-be-explained outcome. Within this perspective, the goal is to design ABMs that are "high fidelity models" (de Marchi and Page 2014), i.e. ABMs that are characterized by "high-dimension realism" with respect to both the model's micro-level assumptions and the macroscopic target one wants to reproduce (Bruch and Atwell 2015).

The KISS and KIDS principles should only be seen as two ideal-typical model-building strategies. As we will see below (Section 3.4), specific ABMs will be closer to one of the two extremes without being perfectly described by any of the two ideal-types. However, the distinction has a heuristic value in that it helps us to see and characterize the large heterogeneity of the ABM field in terms of modeling styles. My intention is not to defend one style against another. An ABM can be designed, analyzed, and exploited with various goals in mind (for a typology of the ABM's goals, see Edmonds et al. 2019). Depending on the goal, different (combinations of) model-building strategies can be helpful (see, again, Edmonds et al. 2019: appendix). ABMs designed and studied along the KISS lines with the goal of finding minimal, counter-intuitive conditions for the emergence of a given phenomenon clearly are a legitimate and important approach for theory construction (see Flache and de Matos Fernandes 2021; Flache et al. 2022). In this book, my focus is on whether, and if so how, an ABM can contribute to defending persuasive causal claims. For this specific goal, I will argue (in Chapter 4), overtly over-simplified ABMs along the KISS lines are insufficient.

3.2 Abstract *versus* Data-driven ABMs: Recent Trends

Although the contrast between Hägerstrand's (1965) and Schelling's (1971) models pioneering ABM methodology in the 1960s shows that this technique can be employed on the basis of very different combinations of theoretical simplifications and empirical data, various literature reviews clearly document that the field has subsequently been dominated by abstract ABMs rather than empirically oriented ABMs (see Macy and Willer 2002; Sawyer 2004; and, more recently, Squazzoni 2012: chs. 2–3).

Arguably, that highly stylized ABMs are still much more frequent than data-driven models is related to how the ABM was framed by the first programmatic work aiming to introduce this method into the social sciences. In Epstein and Axtell's (1996: 4, 18, 20, 22, 177) foundational *Growing Artificial Societies*, for instance, ABMs are seen as "laboratories" in which as simple as possible micro-level rules are shown to be sufficient to generate a macro-level outcome of interest. With the exception of the Sugarscape variant in which agents are supposed to act as neoclassical rational consumers, Epstein and Axtell never use existing sociological or psychological theories to set up these micro-level rules. Similarly, it is only in the book's conclusion that empirical data are considered as a possible source for the design of a "physically realistic environmental model" (*ibid.*: 164). As to the outcome to be explained, Epstein and Axtell put forward the principle of "qualitative similarity" between the simulated outcomes and the real-world macroscopic target but their targets are never precisely defined on the basis of quantitative data or historical cases. Hence, Sugarscape's capacity to generate realistic income distributions, evolving friendship networks, migration dynamics, or a "proto-story", just to name a few collective outcomes the authors are interested in, is difficult to ascertain. Between their simulated phenomena and their real-world counterparts there is only a "phenomenological" analogy. This is the same "puzzling imprecision" that Sugden (2009: 4) already described with respect to Schelling's *ante litteram* ABM of residential segregation—and more generally about "highly abstract models" in economics (see Sugden 2000).

Epstein and Axtell's book had a considerable impact on scholars who started to be attracted by ABMs. For now, *Growing Artificial Societies* received more than 6000 citations (Google scholar count, March 2021). As a rough comparison, this corresponds to about one tenth of the citations currently received by Granovetter's first article on "The strength of weak ties", one of the most quoted journal articles in the social sciences. If one considers that this paper was published more than 20 years before Epstein and Axtell's book, and that Granovetter's paper clearly spanned a much larger scientific audience (Keuchenius et al. 2021), then it appears that *Growing Artificial Societies*, and the KISS principle behind it, arguably had a huge influence on how ABM scholars understood this method.

Many research areas, including cooperation (Axelrod 1997), trust and reputation (for an overview, see Pinyol and Sabater-Mir 2013), the emergence of norms (Axtell et al. 2006), and cultural and opinions dynamics (for an overview, see Xia et al. 2011), indeed continue to be dominated by the Sugarscape style. Deffuant et al.'s (2003) response to some critics of this orientation provides an especially clear illustration of the motivation animating the supporters of the Sugarscape style. In particular, Deffuant and colleagues overtly argued that micro-level assumptions relying on "common sense psychological observations" are not only "beautiful" but also "necessary" because, in their view, sociological and psychological knowledge is still largely theoretically inconsistent across different paradigms and weakly supported empirically.

However, over the last 15 years or so, concordant signs of a moving-away trend from common-sense-based ABMs appeared. Epstein's *Growing Artificial Societies* follow-up book, namely *Generative Social Science* (2006: 12–16)—another best-seller in the field, which has so far collected, on average, more than 100 citations per year (Google scholar count, March 2021)—overtly devoted an entire section to data-driven ABMs, which focuses on clearly defined empirical collective phenomena, and provides a detailed account of research in archeology based on high-fidelity agent-based simulations. Modelers who were used to appreciating the KISS principle began to recognize the importance of using empirically calibrated and validated ABMs (Boero and Squazzoni 2005); some explicitly argued for taking inspiration from more traditional micro-simulation models and using empirical distributions instead of arbitrary (typically, uniform) probability distributions to initialize the agents' core variables (see Hassan et al. 2010). Methods for the importation of real-world social network data within ABMs have also started to be explicitly formalized (see, in particular, Smith and Burow 2020).

Dissatisfaction with abstract ABMs also appeared within research subfields in which the KISS principle has traditionally dominated model-building strategies. For instance, Sobkowicz (2009) extensively reviewed opinion dynamics models and criticized socio-physicists for ignoring the existing sociological/psychological literature in the model-building stage and for virtually never confronting the models' outcomes with clearly delimited macroscopic quantitative data (see also Castellano et al. 2009; ChattoeBrown 2014). Mäs and Flache (2013) were sensitive to these objections and proposed a new model of opinion dynamics where assumptions on individuals' behaviors and interactions are systematically based on empirical research in social psychology on attitude changes and memory processes. These assumptions as well as the opinion trends predicted by the model are then experimentally tested. Flache et al. (2017) have given a programmatic flavor to this orientation. They published a new manifesto for the next generation of studies of opinion dynamics models, where it is explicitly claimed that "the assumptions and predictions of theoretical models need to be put to the empirical test", and strategies for using experimental and survey data to this aim are discussed.

ABMs in which experimental data are exploited to establish the agent's behavioral rules are starting to complement abstract ABMs also in other research areas, including the diffusion of innovations (Cointet and Roth 2007), reputation dynamics (Boero et al. 2010), and game-theoretic ABMs of cooperative behavior (Wunder et al. 2013). Even in economics, where ABMs were initially used only as toy models with no or little use of theory and data, it is now maintained that, for ABMs to go beyond just-so stories, more interaction with experimental research on economic decision making (Duffy 2006) and empirical validation (Fagiolo et al. 2007) are necessary. Recent applications to the study of goods as well as job markets provide concrete research examples of these programmatic statements (see, respectively, Holm et al. 2018 and Kant et al. 2020).

Thus, in agreement with Heckbert et al.'s (2010: 45–6) review of ABMs at the intersection of ecology and economics, it seems descriptively fair to conclude that, although "the calibration and the validation of models" remains one research frontier for the field, "ABM has increasingly moved from exploratory models with *ad hoc* representations of underlying processes to face the rigor of empirical validation".

3.3 Theoretical, Input, and Output Realism

If the literature I have just considered documented increasing efforts to achieve more realistic ABMs, a careful inspection of this literature also suggests that this trend actually takes different forms. In particular, three tendencies can be detected:

1) a trend towards what I propose to call "theoretical realism", i.e. the tendency of replacing ABMs whose assumptions on the entities' behavior of interest are based on common sense and intuitions with ABMs where these micro-level assumptions are informed by clearly explicated sociological and/or psychological theories (possibly supported by empirical and/or experimental evidence);

2) a trend towards what I propose to call "input realism", i.e. the tendency of replacing ABMs whose parameters and functions are based on arbitrary choices of statistical distributions and functional forms with ABMs where parameters and functions are estimated through empirical and/or experimental data—an operation that I will hereafter generically call "empirical calibration";

3) a trend towards what I propose to call "output realism", i.e. the tendency of replacing ABMs that only produce qualitative abstract outcomes with ABMs where the simulated outputs are systematically confronted with well-specified datasets describing the target of interest—an operation that I will hereafter generically call "empirical validation".

The three trends are clearly related to a larger diffusion of the KIDS principle introduced above, which essentially asks for as much empirical evidence as possible to be exploited when building an ABM. In addition to this general "force", one can speculate that the trend towards "theoretical realism" is more related to increasing awareness of the rule of thumb that Edmonds et al. (2019: appendix) called "Enhance the Realism of the Simulation" (EROS), which essentially concerns a requirement of higher psychological plausibility of an ABM's micro-level. Finally, the trend towards "input realism" is probably related to a larger diffusion of the rule of thumb that Edmonds et al. (2019: appendix) called "Minimize the Number of Free Parameters" (MNFP).

In what follows, I interpret "theoretical", "input", and "output" realism as different ingredients of the general level of realism a given ABM can achieve. By "realism", I intend the extent to which the model models the mechanisms of interest in a way that reflects what is known on these mechanisms so that the model can be considered a "credible" representation of the mechanisms. As to how "credibility" can be built, I am thus following here the simple principle proposed by Sugden (2000: 23) according to which "if we are to make inductive inferences from the world of a model to the real world, we must recognize some significant similarity between those two worlds". I see "theoretical", "input", and "output" realism as three ways of increasing the degree of "parallelism" between an ABM and the real-world mechanism the ABM wants to describe. Obviously, on each dimension, the degree of realism a given ABM can reach should be seen as a continuum, and it seems inevitable that different observers will provide different assessments of the ABM's realism. But an emphasis on the importance of "theoretical", "input", and "outcome" realism is likely to change our views on the extent to which it is acceptable only to rely on subjective experiences, anecdotal observations, stories, common-sense beliefs, and/or qualitative appreciation on how an ABM generates the target of interest. In other words, by looking at a given ABM from the point of view of how it deals with "theoretical", "input", and "outcome" realism, we put ourselves in the best position for countering the general problem that Sugden (2009: 25) recognized in highly abstract models, namely "that authors typically say very little about how their models relate to the real world".

To my argument this point is crucial. If an ABM can indeed be questioned as to its capacity to reflect realistically the mechanism it wants to model, the possibility that the ABM can help causal inference on this mechanistic ground will also be correspondingly questioned. For this reason, it is important to distinguish these different dimensions of how "realism" is currently achieved within an ABM. Since "theoretical", "input", and "outcome" realism bear in different ways on ABM's capacity to contribute to the construction of persuasive causal claims, critics of ABMs may reach different conclusions if, although accepting to inspect more "realistic" ABMs rather than highly abstract models, they still base their analysis on this or that specific ABM. Indeed, as I will show next, the three paths to more "realistic" ABMs have been relatively independent in the sense that they do not necessarily coexist within all ABMs that have the ambition to be more than formalized abstractions for the exploration of theoretical hypotheses (for a discussion of this specific ABM's goal, see Edmonds et al. 2019: 5.1–5.12).

3.4 Different Paths to More Realistic ABMs

As I have noted above, the first systematic treatment of the ABM as a general formal modeling tool for the social sciences—i.e. Epstein and Axtell's (1996) *Growing Artificial Societies*—framed the ABM through the KISS principle (see Colander et al. 2004: 257–8). The series of Sugarscape models designed in this book can thus be regarded as exemplar combinations of low "theoretical realism" and virtually no "input" and "output" realism—both empirical calibration and validation being absent. For this reason, this type of ABM has often been criticized for being only "just-so stories" (or "toy models") in the sense that such models can at best have heuristic value but cannot support statements about real-world mechanisms (for a recent discussion in archeology, see Drost and Vander Linden 2018).

To a large extent, the recent history of ABM research can be described as an attempt to design more realistic ABMs through various combinations of "theoretical", "input", and "output" realism. To document this heterogeneity, Table 3.1 classifies a selection of 23 ABMs from economics, epidemiology, ethno-archeology, demography, marketing research, and sociology according to the level of "theoretical", "input", and "output" realism that is present in each model. By taking Epstein and Axtell's Sugarscape models as a benchmark (Category 1), Table 3.1 synthetically shows that these ABMs exhibit six different ways of combining "theoretical", "input", and "output" realism. Moreover, by considering the extent to which "theoretical realism" is present, Table 3.1 shows that ABMs falling in Categories 2–7 can in turn schematically be organized in two sub-groups that I will respectively call "theoretically blind" and "theoretically informed" data-driven models.[2]

Before I discuss each type in more detail, three qualifications are needed. First, my ABM review is not intended to be exhaustive. I only claim that it is sufficient to illustrate typical combinations of ABMs' "theoretical", "input", and "output" realism, thereby helping to show how the lack of one or more between the three ingredients may provide an additional reason why the use of ABMs for causal inference is often viewed with

---

[2] Although logically possible, I have found no examples of a seventh type, which would combine specific sociological/psychological theories informing the model's micro-level assumptions (theoretical realism) and full empirical calibration without empirical validation (see Table 3.1, Category 8). I suspect that this can be explained by the fact that when one takes pain to build a very realistic model on the input side; this is because one intends to demonstrate how the model fully accounts for some phenomenon, which requires empirical validation as well.

skepticism. Second, although Table 3.1 is depicted in dichotomous terms for visual reasons, "theoretical", "input", and "output" realism should obviously be considered as a continuous dimension. My point here is to document the variability of their possible combinations, for which a dichotomous simplification seems sufficient. As a consequence, the summary in Table 3.1 should only be seen as a map providing a benchmark to locate a given ABM within the complex space defined by the three dimensions. Finally, compared to other categorizations of ABMs that are available in the literature (see, for instance, Boero and Squazzoni 2005; Brenner and Werker 2007), mine differs from existing ones in that my ultimate goal is to propose a categorization of ABMs that helps us judge to what extent ABMs can contribute to causal inference on mechanistic grounds. For this reason, the typology of Table 3.1 is built in such a way that it allows us to identify categories that are sufficiently fine-grained as to make it possible to compare ABMs to concurrent methods for causal inference (which will be done in Chapter 5).

3.4.1 "Theoretically Blind" Data-driven ABMs

The strong impact that the KISS principle had on ABMs is testified by the fact that even ABMs seeking to go beyond "toy models" may score very low on "theoretical realism". By this I mean that the micro-level assumptions adopted to design the model are not systematically grounded in existing sociological and/or psychological theories. The modeler remains instead for the least agnostic about the possibility that the rules agents follow to

Table 3.1 Observed combinations of "theoretical", "input" (through empirical calibration), and "output" (through empirical validation) realism in 23 ABMs from economics, epidemiology, ethno-archeology, demography, marketing research, and sociology.

Category	Case studies	Theoretical realism	Input realism	Output realism
1	Epstein and Axtell (1996)	–	–	–
	"Theoretically blind" ABMs			
2	Bruch and Mare (2006); DiMaggio and Garip (2011); Fountain and Stovel (2014); Dugundji and Gulyas (2008); Wunder et al. (2013)	–	×	–
3	Arthur et al. (1997); Lux and Marchesi (1999)	–	–	×
4	Ajelli et al. (2010, 2011); Bruch (2014); Frias-Martinez et al. (2011); Hedström (2005)	–	×	×
	"Theoretically informed" ABMs			
5	Janssen and Jager (2001, 2003); Delre et al. (2010); Manzo and Baldassarri (2015)	×	–	–
6	Todd et al. (2005); Billari et al. (2007); Gonzales-Bailon and Murphy (2013); Silverman et al. (2013); Manzo (2013)	×	–	×
7	Mäs and Flache (2013); Manzo et al. (2018)	×	×	×
8	*Not found*	×	×	–

make their decisions can be deduced from sociological and/or psychological theories that have been discussed, and possibly at least partly empirically validated, within the literature. Realism is rather sought by using empirical data either on the input side (through empirical calibration) or on the output side (through empirical validation). I propose to call this type of ABM "theoretically blind" data-driven models.

ABMs in *economics* that seek to go beyond toy models by maximizing *only* empirical calibration provide a first illustration of what "theoretically blind" data-driven ABMs can look like (see Table 3.1, Category 2).

For instance, Dugundji and Gulyas (2008) studied the individuals' transportation mode in the municipality of Amsterdam by relying on a combination of econometric techniques and agent-based simulations. The ABM was entirely empirically calibrated in that the agents' behavior was based on a series of (a-theoretic) discrete choice models estimated on survey data predicting the individual's choice for the transportation mode as a function of individuals' socio-demographic features (e.g. gender, income, age, education, and residential location) and influence variables reporting the proportion of people living in the same district with similar socio-economic profile choosing this or that transportation mode. While the statistical model provided a cross-sectional picture of the marginal effects of these variables, Dugundji and Gulyas turned to ABMs to embed the statistical models into a dynamic framework so that they could explore the aggregate consequences of the statistically estimated behavior when actors are assumed to be exposed to changes in locally aggregate behaviors. Thus, we have here an ABM in which the agents' behavior is descriptively accurate but no reference to existing micro-level sociological or psychological theories is made. Moreover, while the model is entirely calibrated on the input side, no empirical validation on the output side is performed.

A similar combination of lack of "theoretical realism" at the micro-level, full empirical calibration at that level, and no systematic output empirical validation can be found in research at the intersection of *computer science* and *behavioral economics*. For instance, Wunder et al. (2013) studied cooperation behavior through a variant of public good games in which the final reward was not computed on the entire population of players but only over the player's neighbors. After collecting data from a large-scale web-based experiment, they designed a series of formal models predicting individual contributions and fitted each of them to experimental data. Then, they employed an ABM in which agents behave according to the formal model best fitting the experimental data in order to explore the agents' cooperative behavior under theoretical conditions that were not covered by the original experiment. Wunder et al. described clearly the logic behind their analysis when they claimed: "our approach preserves the 'ABM as thought experiment' tradition of ABM, but attempts to ground it in agent rules that are calibrated to real human behavior within at least some domain". At the same time, they acknowledge that the way they calibrated empirically the agent's behaviors gives priority to predictive accuracy over cognitive plausibility.

In *sociology*, ABMs with similar features can also be found. For instance, DiMaggio and Garip (2011) want to explain the persistence of inequality in adoption rates of a new technology (namely, the Internet) among actors with different educational backgrounds. To this end, they built an ABM in which the agents' choices depended on the economic costs of the new technology and the prior choices of the agent's social contacts. On the input side, the size of the agent population as well as the agents' attributes like income, educa-tion, race, and network size came from US survey data. On the output side, DiMaggio and Garip simulated the model by manipulating the internal composition of the agents' net-work in terms of homophily along income, race, and education lines. Similarly, in their study of the role of network structure and referrals in job career instability, Fountain and Stovel (2014) only calibrated on US empirical data the distribution of the worker's skills and the firms' size. Empirical calibration of the agents' behavioral rules without macro-scopic empirical validation was also clear in Bruch and Mare (2006: 690–4), who exploited US vignette-based survey data for estimating the function that better described the actors' ethnic preferences. Then, they inject the estimated function within the bi-dimensional version of Schelling's original segregation model and explore the levels of racial segrega-tion that emerge. Thus, in all these ABMs, no explicit reference to sociological and/or psychological theories is made to justify the agents' behavior; output realism is also low because the ABM's simulated macro-regularities are not systematically confronted with empirical data; but, various forms of empirical calibration of the agents' core attributes are present.

Some models in *computational finance* illustrate a different version of "theoretically blind" ABMs, where one seeks to take distance from toy models by introducing "output realism" through some form of empirical validation—with no use of specific sociological/psychological theories or attention to empirical calibration on the ABM's input side (see Table 3.1, Category 3).

In these models, the macroscopic phenomena of interest typically are recurrent statisti-cal properties of financial markets such as the fat tails of the unconditional distributions of returns or the volatility clustering and persistence of prices. To generate these statistical features, Lux and Marchesi (1999), for instance, modeled the stock market as a fluid undergoing phase transition, where traders (fundamentalists, pessimist chartists, and optimist chartists) switched from one group to another depending on the comparison of the respective profits, an opinion index, and the price trend. In the so-called Santa Fe artificial stock market, Arthur et al. (1997) endeavored to account for price fluctuations by an evolutionary mechanism, where traders learned from the observation of prices by modifying their trading strategies through random mutation and cross-over of their best performing trading rules. The general hypothesis that both models wanted to test was whether the statistical properties of interest depended on agents' behavioral heterogene-ity, a feature that traditional rational-choice macroeconomic models do not postulate. In both cases, the ABMs' capacity to reproduce the robust statistical features observed in real stock markets over very long time intervals, together with the presence of behavioral het-erogeneity, is taken to compensate the lack of realism of the micro-level assumptions and the scarce attention to input-level calibration.

A last variant of "theoretically blind" ABMs combines instead "input" and "output real-ism" through advanced forms of empirical calibration and validation (see Table 3.1, Category 4). ABMs of disease diffusion in *epidemiology* provide typical illustrations of this (for a programmatic statement from a software engineering viewpoint, see Parker and Epstein 2011).

For instance, Ajelli et al. (2011) simulated a pandemic at the scale of a single country (namely, Italy) through an ABM in which agents are assigned socio-demographic characteristics, geographical locations, and movement probabilities on the basis of a variety of survey, administrative, and census data. Spatial structures at the level of municipalities are also represented as well as physical locations of schools and workplaces. Agents are assumed to meet at random within households, schools, and workplaces whereas dyadic interactions within the general population depend on geographical distances between actors. The model is used to predict the evolution of influence-like disease at several scales (namely, country, census areas, and municipalities) and within different groups of the population. The model's predictions are then systematically confronted with specific data at various levels of aggregation (Ajelli et al. 2010). Frias-Martinez et al. (2011) showed how it is possible to achieve even more fine-grained empirical calibration by using cell phone records instead of census and survey data. In this way, they argued, interaction probabilities could be based on the real networks of contacts and mobility flows between actors. They calibrated and tested the model on Mexican data concerning the H1N1 flu outbreak in 2009 in order to assess the impact of government preventive interventions. Thus, in these ABMs of disease diffusion, contextual, spatial, individual- and interaction-level empirical calibration was combined with direct forms of outcome validation. However, despite this high "input" and "output" realism, since diseases are supposed to flow from one individual to another without the intervention of any cognitive mechanisms, these models typically do not make any reference to specific sociological or psychological theories of human decision-making, thus keeping "theoretical realism" at the micro-level close to zero.

Similar combinations of empirical calibration and validation, without theoretical realism, can also be found in *sociological* ABMs. For instance, Hedström (2005: ch. 6) studied unemployment in the Stockholm metropolitan area among 20–24-year-old youngsters. In particular, he first estimated a statistical model predicting the likelihood of leaving unemployment depending on actor-level attributes and unemployment among peers, and then made his virtual agents choose to leave unemployment according to the estimated logistic equations. He employed the model with a "counterfactual purpose" (*ibid.*: 138) in the sense that he wanted to assess the extent to which the evolution of the unemployment level in the population of interest changed compared to the actual level (thus, empirical confrontation between simulated and empirical aggregate data was also present) when the value of the coefficient expressing the marginal effect of unemployment among someone's peers was manipulated.

Bruch (2014) followed a similar logic in her study of the impact of changing income inequality between and within race group on the levels of residential segregation. To this aim, she first relied on panel data to estimate discrete-choice models of residential mobility, which she then exploited to calibrate residential choices of virtual agents within ABMs with increasingly realistic geographic constraints and population structures (namely, for three US cities such as Los Angeles, Atlanta, and Chicago). The empirically calibrated ABMs were used to study the aggregate consequences of agents' choices on the level of residential segregation under different levels of between- and within-group income inequality (which thus is the variable that is counterfactually manipulated). Finally, in order to prove that the ABMs' simulated aggregate consequences are realistic, Bruch estimated fixed-effect regression models suggesting that empirical data (for the 100 largest US metropolitan areas) presented statistical features that were in line with those characterizing the previously simulated outcomes. Again, the problem of knowing whether it is realistic to depict real actors' choices as following the assumptions embedded within a conditional logit model for discrete choices is not discussed so that, while empirical calibration and validation are present, "theoretical realism" is not an explicit goal of the ABM.

I do not ignore that one can find arguments justifying why keeping low "theoretical realism" at the micro-level may even be a suitable feature of an ABM. For instance, Hedström (2021) have recently systematized the intuition that the difficulty of accessing fine-gained data on actors' reasons and intentions should lead modelers to direct their efforts of empirical calibration to contextual features, in particular in terms of interaction structures, rather than to the actor-related part of an explanatory model. While the rationale behind this argument is clear, the extent to which low "theoretical realism" could diminish the capacity of an ABM to contribute to causal inference on a mechanistic ground should not be ignored. I will be back to this point in the next chapter (see Chapter 4, Section 4.1.1).

3.4.2 "Theoretically Informed" Data-driven ABMs

"Theoretically informed" data-driven ABMs value the goal of relating the assumptions on agent's behaviors to existing sociological and/or psychological theories and/or to empirical information supporting these theories. Through this strategy, the modeler seeks to give a specific meaning to the rules agents follow to make their choices and the ways these choices are related to structural constraints. As long as the sociological and/or psychological theories at hand have also received some empirical support, the model is regarded as not only meaningful at the micro-level but also realistic as this level. Similarly to "theoretically blind" data-driven ABMs, "theoretically informed" ones come with different degrees of "input" (through empirical calibration) and "output" (through empirical validation) realism.

Some ABMs within *marketing research* provide typical illustrations of "theoretically informed" ABMs without any form of empirical calibration or validation (see Table 3.1, Category 5).

For instance, Janssen and Jager (2001) asked for more realistic ABMs of the diffusion of market products. They argued that, to achieve this goal, it is necessary to use specific psychological theories of preference changes. Consistently, they built on a variety of approaches in social psychology to design the agents' behavioral rules. However, the theory is not used to specify the functional forms of these rules. All the ABM's parameters and agents' variables in the end come from theoretical probability distributions (thus, empirical calibration is absent). Janssen and Jager (2003) attempted to further increase the realism of the model with respect to its network components. Reference is made to the literature on complex networks, which describes what real networks look like, but only abstract topologies were then implemented. As to the target side, the plea for theoretical and empirical realism at the micro-level is not accompanied by a request for empirical validation at the aggregate level. To defend the empirical relevance of the simulated market dynamics, the authors relied only on phenomenological analogies between the simulated outcomes and highly abstract empirical dynamics.[3] Delre et al. (2010) continued Janssen and Jager's original search for psychological and relational realism but the gap between this theoretical constraint and the concrete way the ABM is implemented and validated remained.

In *sociology*, Manzo and Baldassarri (2015) provided another example of "theoretically informed" ABMs with no empirical calibration nor validation. They aimed to explain inequality growth in status distributions, and to this end built an ABM of deference exchanges. Programmatically, they asked for more realistic micro-level specifications of individual behaviors and interactions. In order to fulfill this requirement, the agents' behavioral rules were designed to incorporate existing theoretical and experimental evidence in social psychology and sociology on imitation, homophily, and reciprocity. However, this quest for realism at the micro-level was not accompanied by a direct empirical calibration of the model's lower-level assumptions. Moreover, although recurrent properties of status hierarchies were described at the outset of the analysis as *explananda*, the simulation outcomes in terms of status inequality, though precisely measured, were not directly confronted with macroscopic empirical patterns of status inequality.

Foundational papers in computational *demography* illustrate a second variant of "theoretically informed" ABMs where theoretical realism comes along with "output realism" through some forms of empirical validation but empirical calibration is absent; thus "input" realism is low (see Table 3.1, Category 6).

For instance, Todd et al. (2005) wanted to explain the similarity of the age distributions at first marriage across countries. To this end, they designed an ABM of sequential mate search. Experimental research on cognitive heuristics, which, they argued, provided a more realistic portrait of the actors' choice under uncertainty than traditional rational-choice models, is used to design the agents' behavioral rules. However, similarly to the aforementioned work on market dynamics and status hierarchies, this connection of the ABM's micro-level with the existing theoretical and experimental literature did not not translate into an empirical calibration of the parameters and functional forms specifying the agents' rules. Despite this, and differently from the three previous examples, Todd et al. assessed the relative explicative power of each of their model variants by (qualitatively) comparing the simulated distributions with empirical distributions (namely, for Norway and Romania). The lack of empirical calibration of the model's theoretically motivated lower-level assumptions is acknowledged by Billari et al. (2007) who, despite this limitation, introduced an additional heuristic in the model, namely the imitation of the neighbors' marriage choices. In order to apply this ABM to a specific phenomenon (i.e. population growth in the UK), Silverman et al. (2013) injected detailed contextual demographic data into it but they did not calibrate the model's core micro-level assumptions. They programmatically defended this combination as a means to build a more theoretically oriented demography. Within historical demography, Gonzales-Bailon and Murphy (2013) applied the same strategy to the analysis of long trends in fertility rates in France.

---

[3] This strategy is well exemplified by sentences like "such market resembles the daily shopping of most people, and refers to products such as coffee, toothpaste and milk" (Janssen and Jager 2001: 760).

"Theoretically informed" ABMs with high "output realism" achieved through empirical calibration but low "input realism" because of partial empirical calibration can also be found in *sociology*. For instance, Manzo (2013) focused on specific cross-sectional patterns of inter-generational educational mobility (in France) and built an ABM combining actors' cost–benefit rational evaluations with network-based mimetic behaviors. Here, too, although the model's micro-specification were systematically built on well-identified and partially empirically supported sociological theories of educational choices (theoretical realism), empirical calibration at the input levels was absent (except for the size of agent population and education groups). The simulated patterns of educational inequalities, however, were systematically compared to the empirical ones, and the observed difference between the two series of data was quantified; thus empirical validation was present.

The last variant of "theoretically informed" ABMs that I was able to identify is characterized by the explicit attempt to combine simultaneously "theoretical", "input", and "output" realism (see Table 3.1, Category 7).

A case study approaching this condition is published by Mäs and Flache (2013) who studied opinion bipolarization in small groups. Here, the ABM's micro-level assumptions were deduced from theoretical and empirical research in social psychology on attitude changes and memory processes (theoretical realism); then, a tailor-made experiment was designed to test directly the core behavior postulated for the artificial agents; subsequently, the ABM was partly modified with respect to its micro-level assumptions to make it consistent with the experimental design and some of the model parameters were calibrated through experimental data (empirical calibration); finally, the simulated aggregate predictions on opinion trends were compared with the group-level experimental results (empirical validation).

A study of an ethno-archeological process provided another illustration of an attempt to couple theoretical realism with both "input" and "output" realism through empirical calibration and validation respectively. In particular, Manzo et al. (2018) studied two populations of potters still active in northern India and central Kenya and attempted to explain why, in both communities, innovations (a new firing technique and a new pot shape, respectively) spread faster and more widely in one of the two religious sub-groups present in the two regions. To understand whether the structure of family ties within these religious sub-groups may account for the observed macroscopic differences in the diffusion rates, Manzo et al. designed an ABM whose micro-level assumptions relied on existing theories of how decisions are made when uncertainty is present and complex learning processes are required; then, crucial features of the model were empirically calibrated, in particular the kinship networks through which the diffusion process was supposed to flow; finally, simulated diffusion curves were generated as a function of different versions of network-based choices, and systematic confrontation with the actual diffusion curves was performed. The counterfactual manipulation of the choice mechanism allowed Manzo et al. to show that, while certain network features were necessary to reproduce the by-group diffusion curves observed in India and Kenya, these same features could in fact lead to fast or slow diffusion depending on the quality of the signal (i.e. the behavior of central potters) circulating within the network.

In sum, ABMs across many disciplines are highly diverse. The literature that I examined in this chapter shows that, since the beginning of its history in social science in the

1960s, the ABM was torn between following the KISS and KIDS principles. Although historically the KISS principle had a clear impact on the field, leading to the multiplication of highly abstract models—often generically called "toy models"—, a move away from this type of ABM is clearly in progress. As documented by recent reviews of the literature (see Bianchi and Squazzoni 2015: 299–300, table 3.2), ABMs seeking "realism" are still a minority, but their frequency is increasing.

My meta-analysis of 23 ABMs from economics, epidemiology, ethno-archeology, demography, marketing research, and sociology allowed me to make a more specific point. As surprising as this may seem, ABMs seeking closer links with the "real world" in fact pursued this goal through different combinations of "theoretical realism" (meaning: the attempt of anchoring the agents' postulated behaviors to existing sociological and/or psychological well-defined theories, and/or to evidence supporting them), "input realism" achieved through empirical calibration (meaning: the use of empirical/experimental data to initialize the model's parameters and functional forms), and "output realism" achieved through empirical validation (meaning: the use of empirical information to describe the extent to which the simulated outcomes match the target the ABM intends to reproduce).

I maintain that this observation is key to assessing the sense in which, and the conditions under which, if any, ABMs can contribute to the construction of persuasive causal claims. Given the diversity I have documented, an indiscriminate claim against this method's capacity to contribute to causal inference appears more difficult to defend. Depending on how an ABM is built, studied, and validated, I will argue, it will get more or less close to the goal of being helpful for causal reasoning. Focusing on this or that application, however, would equally lead to biased conclusions because these would depend on the specific limitations of the case study considered. For this reason, I believe that a proper assessment of the ABM's relevance for causal inference should rely on a more abstract discussion of the consequences for causal inference of different degrees of "theoretical", "input", and "output" realism within an ABM. This is the task of the next chapter where I will try to defend the claim that, the more the three forms of realism are simultaneously present within an ABM, the more an ABM can contribute to causal analysis (from a production perspective, in Chapter 1's terminology).

Part 2

Data and Arguments in Causal Inference

4

Agent-based Models and Causal Inference

In what sense and *under which conditions*, if any, can agent-based models (ABMs) contribute to causal inference? This is the central question of this book. In the previous chapters I have accumulated the conceptual and methodological elements that seem to me necessary to provide a principled answer to this question.

In particular, in Chapter 1 I started with defining causal inference as the set of cognitive operations through which, from a limited set of empirical and theoretical information, one tries to argue, on the basis of different types of methods, data, and arguments, that one specific happening systematically alters the probability that another happening follows. I then showed that the concept of causality, which enters this definition, as well as that of mechanism, which is employed by different methodological approaches to causal inference, have in fact received different accounts. Without acknowledging this diversity of views, I argued, the question of the extent to which ABMs can contribute to causal inference cannot be properly answered. The answer to this question indeed is likely to be contingent on the views on causality and mechanism one implicitly endorses.

Chapter 2 added a first set of methodological clarifications. It explained why the technical infrastructure of an ABM makes this method square with a production account of causality and a vertical view of mechanism. This is because an ABM is such that it can perform two tasks: first, it can generate the sequence of events that a set of low-level (or small-scale) units in interaction are able to trigger; and, second, it can progressively generate the outcome associated with this sequence of events. Once this is understood, one is on the right track to being able to answer the "in what sense" part of my research question: an ABM can be relevant for causal inference *in the sense* that it provides a tool to model explicitly competing mechanisms (vertical sense) potentially responsible for the given dependence relationship of interest. In Chapter 6 I will clarify why modeling these mechanisms explicitly is a necessary ingredient to build persuasive causal claims. For now let us assume that this is the case.

At that stage of the analysis, however, the "under which condition" part of my question remains to be addressed. Indeed, even accepting the specific point of view of causality and mechanism embedded into an ABM, one could still object that, although an ABM can *always* detail the source of a given connection among (a set of) variables within the closed world of the model, the ABM may be less strong in generating data and arguments convincing a given audience that the ABM realistically mirrors the relevant mechanisms

Agent-based Models and Causal Inference, First Edition. Gianluca Manzo.
© 2022 John Wiley & Sons, Inc. Published 2022 by John Wiley & Sons, Inc.

underlying the connection of interest in the real world. In other words, according to this objection, what an ABM could achieve *within* the computer would be qualitatively different from what the ABM can teach us with respect to what happens outside the computer. I admitted this point myself in Chapter 3 where I accumulated a second set of methodological elements on ABMs by documenting how ABMs in fact greatly differ regarding the way the modelers anchor the mechanism of interest to specific theories and make this mechanism communicate with real-world data describing the model's target. In particular, I showed that ABMs are recurrently built on different combinations of "theoretical", "input", and "output" realism (see Chapter 3, Table 3.1).

Building on these conceptual and methodological elements, I now first explain the *ideal conditions* under which an ABM can contribute to causal inference from a production point of view on causality and a vertical perspective on mechanism (Section 4.1); then I describe the *practical limitations* hampering the use of ABMs for causal inference on this mechanistic ground (Section 4.2); and, finally, I discuss *from-within-the-method* solutions to circumvent such limitations (Section 4.3).[1]

4.1 ABMs as Inferential Devices

In this section, I consider each of the three dimensions of the realism of an ABM that I have identified in the previous chapter, i.e. "theoretical", "input", and "output" realism (see Chapter 3, Section 3.2) and discuss how their presence (or absence) impacts on the possibility that an ABM can be used to gain knowledge on generative mechanisms (in the vertical sense) that can be extrapolated from the world of the model to the real world. This discussion will lead to the identification of the *in principle* conditions under which an ABM can contribute to causal inference on a mechanistic ground.

4.1.1 The Role of "Theoretical Realism"

Let me start by considering the simplest form of "theoretically blind" ABMs, namely "toy models" that have inspired so many applications in the footsteps of Epstein and Axtell's Sugarscape style (see Chapter 3, Table 3.1, Category 1). As we have seen, in this type of ABM the model's assumptions on the agents' rules are only weakly, if at all, anchored to pre-existing sociological and/or psychological theories, or to empirical evidence supporting these theories (lack of theoretical realism). On the other hand, data are used to initialize the parameters' values and the functions designed to implement the agent's behaviors and interactions (lack of empirical calibration leading to low "input" realism). Finally, data are not used either to assess the extent to which the simulated outcomes reproduce the real-world regularities the model aims to generate (lack of empirical validation leading to low output realism).

Typically, toy models are so simple that it is (relatively) easy to fully inspect and understand their internal functioning. They thus produce generative knowledge that can be

---

[1] This chapter builds on and extends Casini and Manzo (2016: 29–38).

interpreted in causal terms from a *production* perspective in the sense that toy models teach us what dynamic chain of events, or processes, connect the model's micro-level assumptions to a given set of macro-level consequences. "Understandability" indeed was one of the features that Sugden (2000) recognized as one of the ingredients that contributes to create the "credibility" of highly abstract models in the style of Schelling's model of residential segregation.

This knowledge may be causally interpreted even from a *dependence* account of causality. Through systematic manipulation of the model's parameters (i.e. sensitivity analysis, which I discuss in Section 4.3.1) and systematic variation of the model's internal components (i.e. robustness analysis, which I discuss in Section 4.3.2), it is indeed possible to establish (probabilistically) that a given parameter/aspect of the model is responsible for a change in macro-level simulated patterns, all other model parameters/aspects being held constant. Given the complete closed nature of the system, it is also possible to ascertain by which modifications of the simulated process this change in the macro-level results was obtained. By the way, it is precisely on this basis that some epidemiologists defended the deep link between ABMs and the potential outcome approach with its counterfactual understanding of causality (see Marshall and Galea 2015).

However, the causally interpretable knowledge produced by highly abstract models is purely internal to the numerical system instantiating the ABM under scrutiny. This is the consequence of the simultaneous lack of theoretical realism and empirical calibration on the model's input side, and empirical validation on the model's output side. As to the former, since no specific theory or data supporting the micro-level assumptions are present, it is impossible to argue that the ABM's low-level infrastructure mimics any well-defined aspect of social reality. As to the latter, since the model's macro-level numerical consequences are not confronted with any clearly identified empirical regularities, it is unclear what the ABM is actually replicating. Thus, the model does not entitle us to claim that the mechanisms depicted by the model parallel the mechanisms at work in the real world. As a consequence, one does not get any causal knowledge, either from a dependence or from a production viewpoint, on any target micro-level mechanisms, or the dynamic process generated by it. Highly abstract ABMs clearly have no direct utility for causal inference.

"Theoretically informed" ABMs (see Chapter 3, Table 3.1, Category 5) constitute a first improvement. In this case, the model's behavioral hypotheses are grounded in specific sociological and/or psychological theories (and/or empirical evidence supporting them). As a consequence, the presence of a certain degree of "theoretical realism" gives a clear meaning to the agent's properties, rules of behavior, and reactions to other agents.

To understand this point, let us go back for a moment to "theoretically blind" ABMs. In this type of ABM, what reason is there to believe that, say, an agent's attribute called "opinion" represents a real-world actor's opinion rather than any another individual-level attribute? Similarly, what reason is there to believe that the functional form expressing the behavior of a consumer is more suited to represent a consumer than the behavior of a different type of real-world entity? As any other formal model, ABMs are intrinsically dynamic numeric systems; thus the labels one puts on this or that agent's attribute or behavior is arbitrary. The presence of a specific theory and/or empirical data supporting it (i.e. "theoretical realism") reduces the arbitrariness by imposing a precise meaning on the numerical and logical symbols on which the ABM operates. This, in turn, makes the results more easily interpretable with respect to the target. For this reason, an observer should consider the generative knowledge produced by a "theoretically informed" ABM as more relevant for causal inference than that generated by some "theoretically blind" ABMs designed using common-sense intuitions on how actors behave and react to each other.

In passing, it is worth noting that a very similar argument was put forward by Willer and Walker (2007: 12, chs. 3–4) with respect to lab experiments, which are usually seen as the gold standard to establish causal claims from within a dependence (namely, of a counterfactual kind) account of causation. In particular, Willer and Walker distinguished "empirically driven" and "theory-driven" experiments, the former being designed on the basis of the method of differences in order to discover new correlations while the latter are designed on the basis of a specific theory in order to test this theory. Willer and Walker's (2007: ch. 6) main argument is that "theory-driven" experiments are not exposed to the limitation of generalizability (i.e. external validity) typically attributed to lab experiments. The reason, they argued, is that "theory is the bridge that connects observations made in the controlled laboratory environments to the world outside the lab" (*ibid.*: 58).

To this argument on how "theoretical realism" is consequential for ABM's potential contribution to causal inference on a mechanistic ground, it may be objected that the gain from using specific micro-level theories as ABMs' inputs still has the limitation that the credibility of these theories is itself contingent on specific audiences, on historical contexts, and on the empirical support they have received at the time they are used to input the ABM under scrutiny (see Kalter and Kroneberg 2014: 108–9). Moreover, sociological and psychological theories are themselves open to interpretation and often portray actors in very different ways. In this sense, micro-level theories alone would be insufficient to make an ABM more realistic so that even "theoretically informed" ABMs would be of limited utility for causal inference.

4.1.2 The Role of "Output Realism" and Empirical Validation

To go a step further, a first solution would consist in augmenting "theoretically informed" ABMs with "output realism" that the modeler can try to achieve through "empirical validation". By this term, I mean the analysis of specific datasets to assess the extent to which the dependence relationships the ABM is supposed to generate are actually present among the macroscopic regularities generated by the simulation of the mechanism implemented within the ABM. This form of empirical validation is sometimes called "output validation", i.e. "checking whether the model generates plausible implications, that is whether the model delivers output data that resembles, in some way, real-world observations" (Delli Gatti et al. 2018: 165).

In Chapter 3 (see Table 3.1, Category 6), I briefly presented several ABMs that systematically compared the macro-level simulated data with well-defined and quantified cross-sectional patterns or historical time series. In this way, these ABMs were able to assess the extent to which the real-world dependence relationships they aimed at explaining were actually produced by the ABMs' postulated micro-level specification. Moreover, by manipulating certain aspects of the ABMs' specification, the confrontation between the simulated outcomes and the empirical target enabled these ABMs to assess how this or that model's manipulation gets the model's output away from the observed regularities. Two reasons explain why "output" realism pursued through "empirical validation" is consequential for the use of an ABM for causal inference.

On the one hand, a clear definition of the real-world target that the simulated outputs is supposed to replicate benefits from the precision of the theoretical meaning of the model's micro-level assumptions: the more specific is the target to be generated, the less likely it is that a loosely defined set of starting assumptions can be used to generate it. On the other hand, in order to claim that a given low-level specification of the ABM is sufficient to generate a given outcome, and that a given intervention on certain components of the ABM's specification changes the probability of observing an outcome, the outcome must be clearly specified. If there is no well-defined and quantified macro-level pattern, the capacity of the ABM's multi-level dynamic to produce this pattern cannot simply be determined. To use a metaphor, it would be like estimating a regression model without measuring the dependent variable.

For these reasons, "output realism" pursued through "empirical validation", coupled with "theoretical realism", increase an ABM's relevance for causal inference because it puts constraints on both the meaning of the model's inputs and the form of the model's outputs, thus ultimately reducing the uncertainty as to the possibility that the connection between the mechanism and the outcome as depicted by the simulation does not correspond to the way this connection is created in the real world.

To this, one may still object that, no matter how far "output realism" will be pushed through "empirical validation" for the ABM under scrutiny, it is always possible to find at least another equally well-specified ABM based on similarly meaningful micro-level assumptions that would be able to generate the real-world dependence relationship of interest with a comparable fit. In this sense, the objection goes, even a "theoretically informed" and empirically calibrated ABM cannot produce data and arguments that are sufficiently convincing because one cannot exclude that different mechanisms postulated by competing theories implemented into concurrent ABMs would be equally good at reproducing the connection under scrutiny. Again the relevance of an ABM for causal inference from a production account of causality and vertical perspective on mechanisms would be undermined.

Although not explicitly referred to ABMs, the following statement by Knight and Winship (2013: 284) contains a typical formulation of this objection:

> Standard practice in mechanism-based analysis typically involves observing an association between a possible cause and effect and the positing mechanisms that could potentially link them. Yet, if explanation involves the identification of causal mechanisms, this approach is insufficient. The possible correspondence between a mechanism and an observed association does not imply causality unless it can be demonstrated that the association could only be due to the hypothesized mechanism.

4.1.3 The Role of "Input Realism" and Empirical Calibration

To address this concern, a third ingredient, i.e. "input realism", can be introduced within an ABM, in combination with "theoretical" and "output" realism (through "empirical validation"). The "input realism" of an ABM can be increased through "empirical calibration". By this term, I mean the action of using empirical data to initialize the values, the distributions, and the functional forms on which an ABM builds at various levels of analysis to implement the mechanisms the ABM wants to model. This task is sometimes called "input validation" (Delli Gatti et al. 2018: 169–72)—within which some distinguish "parameter calibration" from "structural calibration", i.e. the anchoring of specific components of an ABM to data (see, for instance, Vu et al. 2020: §2.17–2.20). Whatever label one prefers, what should be clear is that the form of "input validation" I am invoking here should not be conflated with the procedure that consists in using empirical data on the outcome the model is supposed to generate to approximate the best model parameters that allow one to do so. Unfortunately this approach is also often referred as "empirical calibration" or "estimation", and specific techniques to perform this task have now been proposed (see Delli Gatti et al. 2018: ch. 9; Carrella et al. 2020). The form of "empirical calibration" that I am advocating here is different. It is about using data that are independent, or exogenous, from the ABM under scrutiny so that they can be exploited to help adjudication between possible competing models' specifications.

As shown by several ABMs discussed in the previous chapter (see Table 3.1, Categories 4 and 7; for additional examples, see, respectively, Brown and Robinson 2006 and Magliocca et al. 2014), "empirical calibration" can concern the agents' attributes, the functional forms expressing agents' behavioral rules, and/or the network and geographic locations in which the agents are supposed to be embedded. All these elements can be directly derived from survey, census, or digital data. Using the best data available, empirical calibration thus helps to further reduce the uncertainty that, when only "theoretical realism" and "empirical validation" are present, still surrounds (i) the amount and type of actors' heterogeneity that should be postulated; (ii) the specific rules of behaviors that should be adopted; and (iii) the correct representation of the agents' local environments. By anchoring these aspects at the outset of the simulation through data that suggest what is likely to be realistic, "empirical calibration" thus helps competing mechanisms to be excluded or, when there is agreement on the mechanism at work, several specifications of this mechanism to be adjudicated.

In addition, from the point of view of *dependence* accounts of causality, when such empirical calibration is in place within an ABM, it seems correct to regard the ABM's macro-level consequences, which follow an intervention on an empirically calibrated component of the ABM's micro-level specification, as indicators of counterfactual connections in the real world, because the virtual basis on which this intervention operates, for instance the agents' attribute distributions, behaviors, and/or local context, is a *replica* of the real-world counterpart of these elements. This means that, in response to Diez Roux's (2015: 101) critique of ABMs from a potential outcome perspective (see Introduction, Section 1), it can be argued that it is technically incorrect to claim that in ABMs "everything is counterfactual". As long as the ABM's parameters and micro-level infrastructure are directly based on empirical information, competitive mechanisms can be assessed and non-merely-virtual counterfactual claims can be established.

4.1.4 *In Principle* Conditions for Causally Relevant ABMs

From the previous discussion, I believe that the following conclusion can be legitimately drawn: ABMs can produce data and arguments that are relevant for defending causal claims, and this not only from the perspective on causality that is most in line with the spirit of ABM, i.e. a production view, but also from a dependence view, which, as I discussed in Chapter 2, tends to square with a horizontal understanding of the mechanism-based analysis. In particular, from a *production* perspective, an ABM provides an explicit mechanism under the form of a multi-level dynamic system of interacting units that detail how the dependence relationship of interest was brought about. In other words, the ABM provides an explicit answer to the question "why" we observe the effect of the cause we are interested in. From a *dependence* view of causation, the manipulation of specific aspects of the ABM allows specific "what-if" questions to be answered, thus establishing counterfactual dependences between this or that aspect of the mechanism modeled by the ABM and the real-world dependence relationship the mechanism aims at explaining. However, the previous discussion also suggests that, for the ABM to produce data and arguments that are relevant for causal inference *outside* the model, an ABM must satisfy a demanding requirement: theoretical, input, and output realism, the latter two pursued through empirical calibration and validation, respectively, must be maximized *at the same time*. Each of these ingredients indeed puts specific constraints on the ABM so that, in different ways, all of them contribute to increasing an observer's confidence that the model's results extend beyond the world of the model.

First, "theoretical realism", which, to reiterate, is the use of specific sociological and/or psychological theories (and/or empirical evidence supporting them) to justify the ABM's micro-level assumptions, is necessary to provide the model's micro-level infrastructure with a specific meaning so that the agents' attributes and behaviors are not pure verbal labels open to a variety of interpretations.

Second, "output realism", which, to reiterate, consists in the degree to which the model's macro-level consequences are able to produce the real-world dependence relationships of interest—a degree of proximity assessed through systematic comparison of the model's simulated outcomes with well-defined and quantified real-world datasets (an operation I called "empirical validation")—, is necessary, on the one hand, to restrain even further the set of possible theoretical interpretations for the micro-level of the ABM (the more specific the target to be generated, the less probable that theoretically vague micro-level assumptions can be used to generate it), and, on the other hand, to prove that the postulated micro-specification is capable of generating the specific macro-level outcome of interest.

Third, "input realism", which, to reiterate, amounts to the direct injection of empirical information within the ABM's parameters, distributions, functions, and various substantive components (an operation that I called "empirical calibration"), is necessary to eliminate the possibility that alternative micro- and/or network-level mechanisms, or different specifications of the same mechanisms, are equally able to replicate the macro-outcome of interest, thus ensuring that the initially postulated mechanism(s), with its specific set of parameters and functions, is the most plausible *explanans*.

When all these constraints deriving from the simultaneous presence of "theoretical", "input", and "output" realism are in place, an ABM is able to produce generative knowledge that is relevant for causal inference in that the relationships established within the numerical realm of the model between the posited low-level (small-scale) mechanism(s) and its high-level (large-scale) consequences can be mapped onto their real-world counterparts. In other words, the simultaneous presence of "theoretical", "input", and "output" realism maximizes the likelihood that the ABM acts as a "mimicking" device, and, on this ground, the ABM becomes an inferential device. As stressed by Mary Morgan (2012: 337) with reference to earlier simulation models in macro-economics, namely Orcutt's simulation of the business cycle, "[i]t is this mimicking at two levels that enabled Orcutt's simulation to offer both accounts of the world in the model, and a credible basis for inferences to the real world that the model represents".

4.1.5   Can Data-driven ABMs Produce Information *on Their Own*?

The argument that I have just defended concerning the *in principle* conditions under which an ABM can operate as an "inferential device" was challenged by an anonymous reviewer through a line of reasoning that, in my view, merits a detailed discussion.

In a nutshell, the objection would be that even an ABM simultaneously maximizing "theoretical", "input", and "output" realism cannot be regarded as relevant for causal inference because the data that are used for "empirical calibration" (on the input side) are exogenous to the ABM. As a consequence, the ABM would not have causal value *on its own* because the evidence that is necessary to persuade a given audience that the mechanism modeled by the ABM is realistic, thus making the ABM a reliable inferential device on this mechanistic ground, is not produced by the ABM *itself* but by other empirical sources and methods. For this reason, even the best empirically calibrated ABM cannot produce data and arguments for or against a given causal claim that could not be obtained from observational and/or experimental methods *alone*.

Apart from the fact that this objection implicitly assumes that causal inference essentially is a matter of data —a view that this book tries to contrast (more on this in Chapter 5)—, the objection fails to appreciate the specific point of view on causality and mechanism embedded within an ABM, i.e. a production perspective on causality and a vertical view on mechanism.

In fact, it is indisputable that an ABM cannot produce data on horizontal regularities (i.e. robust correlations). That is precisely why one exploits empirical calibration to achieve "input" realism within an ABM. But, when exogenous empirical data are introduced within the ABM, the ABM becomes an empirically constrained device with *its own* behavior. The new information produced by this ABM constrained on the input side concerns the possible high-level consequences of its empirically calibrated low-level bases. Thus, while the horizontal evidence to calibrate the ABM is obviously external to the ABM, the information that the ABM produces concerning the connection between the low-level empirically grounded mechanisms and the larger-scale patterns associated with these mechanisms is a new kind of information generated by the ABM *on its own*. It is the knowledge concerning the connection between different levels of analysis that is originally produced by an ABM. It is in this sense that, when the ABM is empirically calibrated (theoretically realistic and empirically validated), it can be claimed that it is relevant for causal inference from a *production* perspective.

And, as paradoxical as this may seem, when this *production* perspective is accepted, the generative information concerning the connections across levels of analysis produced by the ABM can even be given a causal meaning from a *dependence* account of causality. In particular, when systematic intervention on an empirically calibrated component of the ABM's micro-level specification is performed, probabilistic relations can be established between the parameter/aspect of the model manipulated and its macro-level consequences. Since the virtual basis on which this intervention operates, i.e. the agents' attribute distributions, behaviors, and local context, is a *replica* of its real-world counterpart, the established probabilistic relations between events concerning different levels of analysis can be regarded as indicators of counterfactual connections across levels of analysis in the real world.

An illustration of this capacity of an ABM to act as an inferential device *on its own* can be found in the most extreme forms of micro-simulation where ABMs are used to study the societal consequences of simple demographic behaviors like fertility choices (for a recent example, see Davis and Lay-Yee 2019). In this type of ABM, individuals' (and/or families') attributes and choices are extensively calibrated through census data and the simulation is used to generate information on the possible macroscopic consequences of these choices at a very large scale (Davis and Lay-Yee 2019: ch. 7). This generative evidence, so to speak, is then used to perform counterfactual manipulations illustrating how different policy interventions may modify the macroscopic dynamics (Davis and Lay-Yee 2019: ch. 10). The crucial point here is that the information concerning the connection between different levels of analysis (say, individuals, interactions within families, and the society at large) that is produced by the simulation was absent from the data that were used to calibrate the model. In this sense, when specific conditions are met, an ABM can produce *on its own* new information that is relevant for causal inference.

4.2 *In Practice* Limitations

The *in principle* conditions that I have identified for an ABM to become a mimicking device, thus supporting causal inference on mechanistic grounds, are demanding. One may object that, *in practice*, these conditions are difficult fully to satisfy. In this section, I take this argument seriously and, first, I discuss the practical obstacles we typically encounter when trying fully to calibrate and validate empirically an ABM with the goal of achieving "input" and "output" realism. Then, I raise the question of whether "partial" rather than "full" "input" and "output" realism may still be an acceptable goal for an ABM that is intended to help causal inference.

4.2.1 ABMs' Granularity and Data Availability

The major difficulty with empirical calibration and validation of an ABM comes from data availability. Indeed the flexibility of ABMs for mechanism design comes at a cost: more granularity demands more fine-grained information to set up the model's components at various levels of analysis (actors, networks, groups, structures, for instance).

As to *empirical calibration*, the consequence is that this operation is likely to remain incomplete. In fact if one carefully looks into the examples of empirically calibrated ABMs presented in Chapter 3 (see Table 3.1, Categories 2, 4, and 7) two facts are evident. On the one hand, some aspects of the models' micro-level specification are more frequently calibrated than others. In particular, population-level parameters (namely, the size of the population and agent subgroups) as well as the agents' attributes always come from empirical data; the agents' behavioral rules are less often empirically calibrated and even less are interaction structures (on the difficulty of finding appropriate network data to be directly injected into ABMs, see Rolfe 2014; but, for in-progress solutions, see Smith and Burow 2020); agents' scheduling (what an agent does and when, basically time) is always speculative. On the other hand, when behavioral rules are empirically calibrated, they always follow functional forms (often regression-like) that are descriptively accurate but behaviorally unrealistic—this is in particular the case of Bruch and Mare (2006); Dugundji and Gulyas (2008); Hedström (2005) (see Chapter 3, Table 3.1, Categories 2 and 4). This limitation results from the fact that the input calibration is based on survey data that do not allow fine-grained description of the mechanisms behind the actors' decisions. But similar problems may arise with experimental data, as illustrated by Wunder et al. (2013) (see Chapter 3, Table 3.1, Category 2), who overtly admitted giving priority to prediction over cognitive plausibility in the way they calibrate the agents' behavior.

The obstacles to empirical calibration loom even larger when one takes the intrinsic dynamic nature of ABMs seriously. From this perspective, it appears that, unless rich longitudinal data are available, proper empirical calibration is structurally impossible. As noted by Hansen and Heckman (1996: 100) with respect to simulation models more generally, it is indeed unclear what the value is of using cross-sectional estimates only at the beginning of a simulated dynamic process.

Data availability also poses problems to *empirical validation*. Here the basic problem is to agree on what "replicating the observed patterns" means. When can simulated and empirical data be regarded as close enough? Obviously a variety of standard approaches exist to assess the fit between two datasets (for an overview, see O'Sullivan and Perry 2013: 211–22). The problem rather is that it is still unclear what the best strategy for an ABM is (for various solutions, see, among others, Thorngate and Edmonds 2013; Thiele et al. 2014). This is crucial not only to ascertain the generative capacity of a given low-level specification but also to assess the extent to which a given manipulation of a certain specification gets us away from the observed outcome, which is key to the counterfactual reasoning that ABMs can in principle support.

One may even wonder whether aggregate data on cross-sectional patterns or longitudinal dynamics are the most relevant types of data to constrain an ABM on the output side, thus increasing the confidence one can have in the mechanisms modeled by the ABM as strong candidates to explain the outcome. As noted by Macy and Flache (2009: 262), a mere input–output mapping does not tell us anything on whether the internal functioning of the ABM represents the real-world mechanism (see also Léon-Medina 2017). After all, the generative capacity of an ABM ultimately resides in the processes that the execution of the computer program describing the postulated mechanism triggered. Thus, should not a proper validation of an ABM go through a comparison between the simulated and the real-world process? In other terms, should not "output" realism be complemented with "process" realism? Obviously, if one endorses this view of the validation of an ABM, according to which the target to mimic would be not only the mechanism potentially behind the aggregate outcome of interest but also the process associated with this mechanism, then the limitations imposed by data availability on an ABM's validation become exorbitant.

As noted by Oreskes et al. (1994: 643) with respect to numerical simulations more generally, data limitations are such that confirmation "is always inherently partial". As an illustration of this, Ajelli et al. (2011), for instance, comparing results from an epidemic meta-population model and an ABM, and despite their huge effort to achieve the best possible empirical calibration, noted: "It is however difficult to state which of the two predictions is the most accurate. On one hand the high level of realism of the ABM should make the prediction reliable. On the other hand this high realism is not free of modeling assumptions (...)"[2] Similarly, despite the presence of empirical information at several levels of analysis and systematic empirical validation, Gonzales-Bailon and Murphy (2013: 136) admitted that "the simulations were not capable of providing causal explanations of fertility behaviour".

Should one then conclude from these *in practice* limitations—which, to reiterate, are not *in principle* limitations—that ABMs are of no value for causal inference if they are theoretically guided but only partially empirically calibrated and validated? To answer this question, it can be useful to understand whether "full" rather than "partial" empirical calibration and validation are reasonable requirements for an ABM and, if not, how one can compensate their lack with resources available from within the ABM methodology. Let me focus on the former point in the next sub-section, and leave the latter for the next section.

4.2.2 ABM's Granularity and Data Embeddedness

Historically ABMs were introduced into the social sciences to model aspects of social phenomena that existing data and techniques were unable to capture (see Bonabeau 2002).

At the beginning of his pioneering study containing an *ante literam* ABM inspired by the KISS principle, Schelling (1971: 147) noted that "the simple mathematics of ratios and mixtures tells us something about what outcomes are logically possible, but tells us little about the behavior that leads to, or that leads away from, particular outcomes". Since then, as I fully discussed in Chapter 2, it is a *leitmotiv* that the strength of an ABM is its capacity to deal with heterogeneity, complex interplay between behaviors and networks, and loops among several levels of analysis, all elements that empirical data (and statistical methods to describe them) are unable to grasp with comparable flexibility. This is the main reason why ABM is traveling across disciplines (in epidemiology, see Auchincloss and Roux 2008; in social psychology, see Smith and Conrey 2007). In an influential article

---

[2] The authors refer in particular to how they modeled movements among municipalities and the probability of getting infected through contact in the general population.

on numerical simulations more generally, Oreskes et al. (1994: 664) make this point very explicitly: "Fundamentally, the reason for modeling is a lack of full access, either in time or space, to the phenomena of interest".

Thus, is it not paradoxical to request full calibration for an ABM? Is there not a contradiction in asking a method whose strength is to model aspects of social mechanisms that available data (and existing mathematical and statistical models) are not able to track to be constrained by those same data?

Ultimately, in order to understand the value of ABMs for causal inference, it is crucial to appreciate that there is a fundamental trade-off for ABMs to be genuinely useful, a trade-off between designing fine-grained mechanisms that connect scarce input data to real-world puzzling patterns and designing data-driven mechanisms by drawing on exhaustive inputs. Since data are *de facto* limited, the more one wants empirically to constrain the simulation, the more one is obliged to reduce the granularity of the mechanism(s) that can be postulated. As noted by Fagiolo et al. (2007: 211–12), the quest for empirically calibrated and validated ABMs contains a potentially conservative stance: it can lead to adapting the type of mechanisms we design to available data, which would lead to under-exploiting the method's potential for mechanism design. The capacity of an ABM to produce data and arguments that are relevant for causal inference in virtue of full empirical calibration and validation is inversely correlated to the granularity of what the ABM can model to contribute to causal inference on a mechanistic ground. The more weight is given to full empirical calibration and validation, the less ABMs can be used to gain insight about those phenomena for which data are missing.

To sum up, the ABM's granularity and the requirement of empirical embeddedness to increase ABMs' input and output realism push in opposite directions. At the same time, empirical calibration (and validation) are legitimate requirements when ABM comes to empirical research with the aim of contributing to causal reasoning. Given this trade-off, is there any way to find, in the presence of data limitation, a balance between the very motivation for using an ABM, namely the granularity and flexibility it allows for mechanism design, and the legitimate requirement of empirical calibration and validation for causal inference? I answer this question in the next section.

4.3 *From-Within-the-Method* Reliability Tools

Within the context of a detailed analysis of spatial simulation models, O'Sullivan and Perry (2013: 226) noted that, although there is still a tendency to give confrontation with empirical data priority in assessing the value of a simulation model, "alternative to data confrontation based on model evaluation and statistical validation have come to the fore over recent years". Given the data limitations obstructing full empirical calibration and validation that I have just highlighted, I share this concern with diversifying the operations needed to build ABMs that want to contribute to causal inference.

In particular, I suggest that the knowledge an ABM produces on the potential mechanism underlying the dependence relationships of interest, thus acting as an inferential device from a production point of view on causality, is stronger as long as, in addition to the best empirical calibration/validation permitted by data availability, the ABM would also systematically undergo a variety of *reliability* checks. The ABM as a methodological framework indeed contains several tools that can be used to assess the variability of a given simulated outcome along a certain number of dimensions when empirical information is insufficient to tell us how this or that part of the ABM should be designed to be realistic. In the present section, I focus on the following tools:

(1) *Sensitivity analysis*—i.e. the assessment of the variability of the ABM's simulated outcomes as a function of the model's parameter values;
(2) *Robustness analysis*—i.e. the assessment of the variability of the ABM's simulated outcomes as a function of the model's internal details;
(3) *Dispersion analysis*—i.e. the analysis of the variability of the ABM's simulated outcomes across repeated simulation of the model under the same set of parameter values;
(4) *Model analysis*—i.e. the analysis of how the ABM works internally once the simulation of the mechanism it models is launched.

Each of these tools should be seen as a complementary resource to empirical calibration (aiming at establishing the ABM's input realism) and validation (aiming at establishing the ABM's output realism) in that, for all the aspects of the model that cannot be anchored to data, these operations in different ways contribute to assessing the dependence of the ABM's macroscopic simulated outcomes on the model's assumptions, thus ultimately helping to reduce the uncertainty generated by data limitation and giving elements to an observer to judge the credibility of the causal inferences that can reasonably be made through the ABM on the mechanism(s) it is supposed to mirror.

Let me clarify that, in what follows, by "model's assumptions" I will refer to every aspect of a given ABM for which data are insufficient to determine what that aspect looks like in the real world. I appreciate that, among "model's assumptions", one may want analytically to distinguish "formal" and "substantive" assumptions. The former would refer to *generic properties of the model* like the type of probability distributions postulated for a given model parameter, the functional forms used to relate some agents' attributes or the topology of the network supposed to connect these agents. In contrast, substantive assumptions would concern *specific elements of the posited core explanatory mechanism* like the agents' action logics or the impact of local interactions on agents' behaviors, typically as derived from domain-specific theories and knowledge. The distinction is obviously a matter of degree, and the specific content of formal and substantive assumptions varies across ABMs. In general, however, formal assumptions are required to implement substantive ones (I will discuss this point further in Chapter 5, Section 5.4.1).

In what follows, I will adopt the general term of "model's assumptions" because, no matter whether formal or substantive, assumptions play the role of making the causal inference possible. Indeed, the stability of the ABM's behavior when the parts of it that *are not* anchored to data, hence the ABM's assumptions, are varied increases an observer's confidence that the macroscopic simulated effects associated with the parts of the ABM that are empirically grounded can be regarded as indicators of empirical connections across levels of analysis in the real world rather than being the result of the model's assumptions. That is why the dependence of the macroscopic simulated outcomes on the model's assumptions, either formal or substantive, must be systematically assessed through reliability tools. Let me now explain each of them in turn.

4.3.1 Sensitivity Analysis

Sensitivity analysis can be performed in different ways (for an overview, see Saltelli 2000; more specifically on ABMs, Thiele et al. 2014; Delli Gatti et al. 2018: 151–62). "Global" sensitivity analysis, whose aim is to fully describe the model's behavior over its entire parameter space, is especially important for two reasons.

First, once the complete simulated input-output mapping is produced, it is easier to discover errors and/or logically inconsistencies in the model. A systematic numerical exploration of the ABM's elements for which data are absent and/or insufficient prevents us from requiring empirical calibration for model's aspects that are inconsistent and/or useless. In this sense, when conceptual exploration of an ABM—an operation usually associated with ABMs approached from a KISS perspective (see Sugden 2000: 8–11)—takes the form of a global sensitivity analysis of the ABM's parameter space, conceptual exploration is also crucially important for ABMs that want to be used to support causal inference on mechanisms outside the world of the model (on this type of loop between KISS-like ABMs and more realistic ones, see Flache and Matos Fernandes 2021). Second, as noted by Helbing (2012: 42), since empirical/experimental estimates that can be used as a model's input are themselves uncertain, knowledge of the ABM's behavior within the model's larger parameter space is important to put in perspective the empirical data used (on this point, for numerical simulations more generally, see also Oreskes et al. 1994: 641–2).

Properly conducted, sensitivity analysis also helps counter the criticism that ABMs lead to results that are unreliable because they are based on specific numerical inputs—which is a common criticism to simulation-based methods more generally (see, for instance, Fararo and Kosaka 1976: 431–3; Sørensen 1976: 85, 89; Gould 2002: 1169–70). When global sensitivity analysis is used to design the ABM's response surface (see, in general, Law 2007: 643–55, and, for an example, Fararo and Butts 1999: 51–2), it becomes more difficult to criticize ABM for not being able to generate fully specified and generalizable results (see Leombruni and Richiardi 2005: 106; Epstein 2006: 29–30; Delli Gatti et al. 2018: 41–2).

Global sensitivity analysis may be difficult, in particular when the dimensionality of the parameter space to be inspected increases. In this case, heuristic-based evolutionary procedures are one possible solution (Stonedahl and Wilensky 2010a, b). Alternatively, sampling-based approaches to "local" sensitivity analysis can be applied (see Saltelli et al. 2000: chs. 2 and 6). Among the ABMs discussed in Chapter 3, Manzo and Baldassarri (2015) relied on this type of sensitivity analysis and, by replicating previous formal models of status hierarchies through ABMs, they showed how proper sensitivity analysis can help to increase one's confidence in the ABM's capacity to identify mechanisms that generate consequences in line with theoretical expectations and background empirical knowledge.

4.3.2 Robustness Analysis

If sensitivity analysis focuses on the values of an ABM's parameters, robustness analysis is directed to the ABM's "internal" cogs and wheels and aims at assessing how varying these cogs and wheels impacts on the ABM's outcomes (Railsback and Grimm 2019: 300, 312–13). Some of an ABM's components that are especially important to manipulate systematically are (see Axtell 2001 and Miller and Page 2004):

(1) The probability distributions from which the value of the agents' properties are drawn;
(2) The functional forms used to relate the agent's properties, thus expressing the agents' behaviors;
(3) The spatial/relational structure through which agents are supposed to interact;
(4) The order in which agents' rules are executed and the agents' scheduling.

As noted above, empirical data are usually missing to fully calibrate all of these aspects, which leads critics to argue that ABMs' results are unreliable because they depend on too many free-to-vary internal details of the simulation (see, for instance, Grüne-Yanoff 2009a: 547).

Robustness analysis is a crucial tool the ABM framework offers precisely to assess the extent to which the macro-level consequences of a given ABM depend on those components of the mechanism(s) the ABM wants to model for which data are insufficient to tell different specifications of the mechanism apart. The more stable the ABM's outputs are against these possible various specifications of the mechanism of interest, the larger our confidence should be in the possibility of using the ABM to make inferences on the operating of this mechanism in the real world. On the basis of this line of reasoning, some even claimed that robustness analysis can confirm hypotheses (Weisberg 2006).

Among the ABMs presented in Chapter 3, Bruch and Mare (2006) provided an illustration of a form of robustness analysis focusing on the agent-level of an ABM, in particular by showing that the amount of spatial segregation generated by Schelling's model in fact depended on the type of function one employs to represent agents' residential choices (see also Van de Rijt et al. 2009; Bruch and Mare 2009). Similarly, Manzo et al. (2018) explored a variety of actors' imitation heuristics and showed that the supposedly causal effect of certain network features on the speed of diffusion of an innovation in fact can take opposite signs depending on the role played within the diffusion process by the most central nodes in the network.

4.3.3 Dispersion Analysis

ABMs typically include various stochastic components. For instance, the agents' attributes are usually initialized from probability distributions; the agents' behavior tends to be probabilistic; the agents' invoking order is randomized. For this reason, one simulation can be seen as a specific realization of an unknown stochastic process. As a consequence, for a given set of initial conditions, the ABM's outcome will be different each time the simulation is run. *Dispersion analysis* concerns the quantification of the variability of the ABM's simulated outcomes when the ABM is simulated several times under the same set of parameter values—hereafter I will call the repeated simulations of the same model "replications". Thus, dispersion analysis is different from "uncertainty analysis", which is about assessing the impact of the potential variability of an ABM's given parameter on the ABM's outcome, and, in this sense, it is a particular form of sensitivity analysis (see Railsback and Grimm 2019: 300, 307–12).

As stressed by Miller and Page (2007: 74–5), critics of ABMs tend to regard this across-replication variability as a lack of predictability in the results generated by an ABM. In fact, when global sensitivity analysis is combined with repeated simulation of the model, a multi-dimensional space can be produced, where for each combination of parameter values a distribution of outcomes is generated. Each distribution is then described by appropriate dispersion measures and the overlap between distributions, hence the potential lack of significance between two manipulations, is evaluated (see Helbing 2012: 47). In this way, an ABM can generate predictions under the form of outcomes with a certain probability.

This procedure is relevant when empirical calibration cannot be fully realized. For an ABM's components for which empirical data are not available or are insufficient, that changes in (a subset of) parameter values do not lead to appreciable differences in the simulated outcomes increases our confidence in the reliability of the ABM's outcomes. As to empirical validation, when appropriate data are available, thinking in terms of distributions of simulated outcomes pushes us to put empirical data in a similar form, for instance by means of re-sampling techniques. This approach would lead to even more rigorous validation of the ABM's micro-level specification, insofar as not only a single observed pattern/trend must be reproduced by the ABM but also the observed (or estimated) variability around it. Among the ABMs commented on in Chapter 3, Manzo (2013) provided an illustration of this type of complex dispersion analysis.

4.3.4 Model Analysis

That ABMs are—or potentially are—black boxes is a common criticism among social scientists (see Morgan and Winship 2014: 341, fn. 15; see also León-Medina 2017). Some philosophers of science even spoke of "epistemic opacity" as a distinctive feature of ABMs (see Humphreys 2009: 618–19). While this critique should be considered seriously, discontents with ABMs seem to underplay the fact that an ABM can be internally inspected as deeply as one wishes. This action can be time-consuming but it is possible. By *model analysis* I precisely refer here to the set of strategies that can be used to understand (and describe) the set of events, behaviors, and feedbacks—in other words, the process—triggered by the mechanisms coded in the ABM once the computer program implemented the ABM is executed.

Apart from sensitivity and robustness analyses, which themselves help us to develop intuitions on which pieces of the model are responsible for the outcomes (see, for example, Railsback and Grimm 2019: 287–91), mathematical techniques based on the theory of Markov processes are available precisely to describe the sequence of states through which an ABM evolves (Young 2006; Izquierdo et al. 2009; Gintis 2013). Among the ABMs presented in Chapter 3, Bruch and Mare (2006, appendix) applied a variant of this approach to their re-analysis of Schelling's model.

The model's main mechanisms can also be introduced sequentially so that it is easier to assess their role in the model's dynamic and their relative weight in the determination of the outcome. Among the models presented Chapter 3, Manzo (2013) followed this strategy when analyzing his ABM of the emergence of educational stratification when a combination of rational and imitation-based choices are present within socio-economic segregated social networks.

Moreover, since an ABM can be simulated again and again, it is possible to collect data on it at various levels of analysis and specific measures can be computed to inspect this or that aspect of the model's dynamic (Railsback and Grimm 2019: 292–3). Among the models presented in Chapter 3, Manzo and Baldassarri (2015) performed this operation to illuminate the link between the model's dynamics at the agent- and the system-level; Fountain and Stovel (2014) followed a similar strategy relying on multi-level regression applied to data collected during the simulation.

Thus, it is always practically possible to shed light on the internal functioning of an ABM, and this should be done carefully and systematically (for a pedagogic and instructive example, see Flache and Matos Fernandes 2021). Understanding how an ABM works is crucially important to assess the value of the ABM for causal inference: knowing how the ABM generates a given (set of) macro-level observed regularities, rather than only knowing whether it is able to do so, provides us with more elements to evaluate the plausibility of the "story" the ABM tells us about the potential explanatory mechanism, thus ultimately helping us to collect arguments to judge if the model describes a process that could have happened in the real world (on this source of a model's credibility, see Sudgen 2000: 25–7).

In sum, in the light of the existence of the above forms of reliability checks, it seems fair to conclude that the ABM methodology has internal resources to cope with the practical problems related to empirical calibration and validation that I have highlighted in Section 4.2. Comparing the simulated outcomes of an ABM to actual data, thus increasing the ABM's "output" realism through empirical validation, is without any doubt necessary to establish whether or not the mechanism designed by the ABM is able to generate the observed outcomes of interest. Feeding empirical information within an ABM is without any doubt also necessary to increase the ABM's "input" realism through empirical calibration and to reduce the probability that several micro-level specifications equally well replicate the patterns/trends of interest. Empirical data, however, are likely to be incomplete. In different ways, *sensitivity*, *robustness*, *dispersion*, and *model* analysis help to assess the uncertainty surrounding data themselves as well as the model's components for which no data are available at all.

Thus, in the end my claim is that an ABM's capacity to produce data and arguments that are relevant for causal inference on a mechanistic ground relies not only on "theoretical", "input", and "output" realism, the two latter aspects being achieved through empirical calibration and validation respectively, but also on the ABM's systematic reliability checks. Reliability checks and empirical validation/calibration of an ABM are complementary tasks: the strength of one set of operations help to counterbalance the limitations of the other. As observed by Muldoon (2007: 883) with respect to simulation more generally, it is only the combination of these operations that can lead to the construction of "robust simulations", i.e. "simulation that can teach us about the world".

To anticipate, in the next chapter I show that, contrary to common views, an ABM is not different from experimental and statistical methods for causal analysis, which also must confront similar practical challenges related to untested and untestable assumptions for which only reliability checks and arguments, but not data, are available.

5

Causal Inference in Experimental and Observational Methods

At this point of the analysis, it is useful to consider again Diez Roux's (2015) reply to those epidemiologists who argued that an agent-based model (ABM) could be seen as a pure implementation of the potential outcome approach, and that, on this ground, this method should be regarded as a tool that can establish counterfactual claims of the same type as experimental and observations methods (see, in particular, Marshall and Galea 2015). I quoted Diez Roux's entire reply in the book's Introduction (see Section 1). Let me focus here on the aspects illustrating how Diez Roux (2015: 101) saw the fundamental differ-ence between ABMs and methods for observational and experimental data that rely on information collected by letting individuals act in real systems:

> (...) The traditional tools of epidemiology are used to extract (hopefully) reasonable conclusions from necessarily partial and incomplete (often messy) observations of the real world. (...) In contrast, when we use the tools of complex systems, we create a virtual world (based on prior knowledge or intuition) and then explore hypothe-ses about causes under the assumptions encoded in this virtual world. (...) In the real world, we have fact (what we observe) and we try to infer the counterfactual condition (what we would have observed if the treatment had been different). In the simulated world, everything is counterfactual in the sense that the world and all possible scenarios are artificially created by the scientist.

That statement illustrates in its purest form the widespread perceived intrinsic superi-ority of observational and experimental methods for causal inference. That these meth-ods rely on "partial and incomplete (often messy) observations" is considered obvious but "hopefully" not too problematic. Interestingly, that observational methods are obliged to rely on assumptions precisely because of data limitation is not even men-tioned. On the other hand, an ABM is seen as entirely relying on "prior knowledge or intuition", which is taken to be "one of the most vexing problems" of the approach; that, as I showed in Chapter 3, an ABM can be anchored to data is not considered. Thus, a "fundamental distinction" is perceived to exist between observational and experimental methods, which deal with "facts", and simulation methods, which are entirely made of "assumptions". The former allow one to establish causal connections in the real world whereas the latter allegedly do not.

Agent-based Models and Causal Inference, First Edition. Gianluca Manzo.
© 2022 John Wiley & Sons, Inc. Published 2022 by John Wiley & Sons, Inc.

In the previous chapter, I demonstrated that this representation of an ABM is not tenable. An ABM can *in principle* couple theoretical realism with "input" and "output" realism through empirical calibration and empirical validation respectively. This combination of ingredients makes an ABM a "mimicking device", thus allowing the modeler to use the knowledge the ABM generated on the connections across different levels of analysis on the basis of a clearly specified mechanism to perform inferences on that mechanism in the world outside the world of the model. When, because of *practical* limitations due to data availability, an ABM's input realism is limited because full empirical calibration is impossible, a variety of reliability tools can be used to quantify the extent to which the ABM's outcomes depend on the modeling assumptions (i.e. every aspect of the model that cannot be based on empirical information, thus acting as a condition for proper causal inference). As a consequence, an ABM that maximizes empirical calibration and validation, coupled with systematic reliability checks, is perfectly able to support causal statements (about real-world connections across levels of analysis) through a combination of data (associated with empirical calibration and validation) and arguments (based on reliability analysis).

In this chapter, my goal is to demonstrate that observational and experimental methods, contrary to widespread perceptions well illustrated by Diez Roux's above-mentioned statement, similarly to ABMs, have to face equally demanding challenges when employed to establish "horizontal" causal claims (in Chapter 1's terminology), i.e. claims about robust probabilistic (from a causes-of-an-effect perspective) or counterfactual dependences (from an effect-of-a-cause perspective). Observational and experimental methods, too, this is the claim I want to make, can formulate credible causal inference on whether changes in an independent variable make a systematic difference to a dependent variable *only* through a mix of data and arguments.

To defend my claim as precisely as possible, I will focus, on the one hand, on randomized experiments (hereafter, in short, RCTs, for "randomized control trials"), and, on the other hand, instrumental variables (hereafter, IVs) and causal graphs (in particular, directed acyclic graphs, hereafter DAGs). This choice is meant to illustrate what challenges traditional methods for causal inference have to face both when confounding can be eliminated by design (RCTs) and when it cannot (IVs and DAGs), in which case other assumptions have to be introduced. Thus, albeit limited, it seems to me that this selection of methods is effective in illustrating the most general issues that are transversal to the main identification strategies of causal effects, i.e. design-based *versus* model-based strategies (see Morgan and Winship 2015: 30–3).

The choice of including DAGs has an additional advantage. It allows the present analysis to be connected to recent studies (in epidemiology) where DAG-informed regression approaches were specifically compared to ABMs, and, again, the contrast was made between the former, which "provide robust estimates of causal effects" on the basis of "assumptions that are transparent", and the latter, for which there is a "lack of consensus about fundamental assumptions or under what circumstances causal effect estimates are valid" (see Arnold et al. 2019: table 1). However, once more, the question of whether (at least) some "assumptions" may be untestable, in practice and/or in principle, in both classes of methods, thus making them equally contingent on data and arguments, was not even mentioned.

Thus, in what follows, after briefly recalling that, in fact, the use of statistical models for causal inference is itself more controversial than it may seem at first (Section 5.1), I systematically scrutinize RCTs, IVs and DAGs more specifically with respect to assumptions that cannot be tested *in practice* because of data availability (Section 5.2), assumptions that are untestable *in principle* because they would require infinite knowledge capabilities (Section 5.3), and how reliability is established *in practice* within these methods (Section 5.4).[1]

5.1 Causal Inference: Cautionary Tales

Statisticians and statistically minded sociologists have recurrently cautioned against simplistic causal interpretations of the empirical estimates produced by statistical methods for observational data. These warnings have concerned a variety of specific techniques among which are path analysis (Freedman 2005), structural equations models (Wang and Sobel 2013), regression for categorical variables (Breen and Karlson 2013), statistical models of network data (VanderWeele and An 2013), and econometric models including externalities based on social interactions (Durlauf and Ioannides 2010).

As Gelman and Hill (2007: 167) observed with regard to regression methods, the common concern here is that "causal interpretations of regression coefficients can only be justified by relying on much stricter assumptions than are needed for predictive inference". A sizable literature also exists on the practical conditions (like sample size) that are required to obtain reliable statistical estimates of causal effects (for a recent statement on the use and misuse of hierarchical regression, for instance, see Bryan and Jenkins 2015). The unrealistic, and often untestable, character of the assumptions allowing causal inference by means of regression-like methods for observational data led some even to propose to limit these methods to descriptive purposes (for an especially radical statement, see Berk 2010; Berk et al. 2013).

However, it seems fair to say that, for now, these cautionary tales have produced relatively limited consequences for research practices on a large scale. It is certainly possible that, as suggested by Morgan and Winship (2015: 13), the "naive usage of regression modelling" is going to disappear. Influential critiques of regression-like models such as those of Andrew Abbott (1988, 1997, 1998) have without a doubt contributed to make the problem visible to a large sociological audience. At the same time, one can suspect that this trend is especially present only within a small elite of researchers (in selected university departments). Review papers in various disciplines assessing how statistical methods for causal inference are applied in empirical research indeed suggest that the overwhelming majority of journal articles do not

---

[1] This chapter builds on and extends Casini and Manzo (2016: section 4). I am indebted to Lorenzo Casini who pushed me to frame the analysis of observational and experimental methods along the same lines as I approached ABMs in the previous chapter. In both cases, indeed, the focus is on what assumptions can, and cannot, be tested, for what reasons this happens, and how the problem can be circumvented. By studying systematically the two families of methods in parallel, the deep similarities between them appear more clearly.

satisfy, or simply ignore, the assumptions required for drawing correct causal inferences (see, for instance, Antonakis et al. 2010). Similar reviews in sociology made the same observation—see, for instance Bollen's (2012: 62–6) analysis of IVs or Breen's (2018) DAG-informed review of studies of social mobility across multiple generations.

The very motivation for pushing the potential outcome approach in sociology seems testifying itself the still low awareness of all technicalities required properly to defend persuasive causal claims. Morgan and Winship (2015: 7, 79, 121) are explicit on this point. On the one hand, they argued that the counterfactual component of the potential outcome perspective can help quantitative sociologists to frame causal questions more rigorously, by importing the experimental protocol within survey-based research; and on the other hand, they saw DAGs as tools to make more rigorous the way data are selected so as to correctly identify causal effects (see also Elwert 2013: 261–2; and, more recently, Makovi and Winship 2021).

By going back to the distinction between "backward" and "forward" causation introduced in Chapter 1—the former focusing *ex post* on the causes of an effect whereas the latter emphasizing, *ex ante*, *what-if* questions centered on the effect of a specific cause (see Gelman 2011)—, it is plausible to think with Morgan and Winship that, as the potential outcome approach to causality involves a shift from backward to forward causal reasoning, the counterfactual view of causation will help quantitative scholars to be more aware of how demanding conditions to establish even a single causal link are. However, the ongoing debate on the "researcher degrees of freedom" clearly shows that the effect-of-a-cause perspective does not constitute *per se* a sufficient protection against interpreting as causal dependence relationships whose empirical ground is questionable (see Simmons et al. 2011). The fact that experiments are at the heart of this debate demonstrates that even causal inferences made through a method that is supposed to represent the best implementation of the potential outcome approach in fact are contingent on a variety of multiple choices concerning data selection, variable coding, case inclusion, comparisons to be performed, and/or inclusions of main or interaction effects (for a detailed list of degrees of freedom, see Wicherts et al. 2016), which ultimately make causal inference contingent on a complex mix of data and always disputable assumptions (see Gelman and Loken 2014).

Compared to this long list of cautionary tales on the extent to which statistical methods can be exploited for causal inference, in this chapter I add an explicit comparison with ABMs and experiments and develop three lines of reasoning of increasing generality. First, I argue that observational and experimental methods, too, have to deal with insufficiently fine-grained data that make some of their assumptions untestable (Section 5.2). Second, I explain how observational and experimental methods, too, rely on assumptions that are, from a methodological point of view, in principle untestable (Section 5.3). Finally, as a consequence of the two first steps, I argue that both ABMs and observational and experimental methods provide a reliable methodology for causal inference only if certain overarching principles do in fact hold (Section 5.4).

Before this, let me reiterate an important qualification. The following discussion has no intention to dismiss observational and experimental methods (in favor of ABMs). My ultimate goal is to motivate my plea for evidential variety (developed in Chapter 6), according to which neither ABMs nor observational/experimental methods *alone* are sufficient to establish persuasive causal claims. The reason, as I will explain later, is that the kind of data and arguments they produce is, from different points of view, equally incomplete and uncertain.

5.2 *In Practice* Untestable Assumptions

As the above-mentioned exchange between Marshall and Galea (2015) and Diez Roux (2015) shows, observational and experimental methods are regarded as intrinsically superior for causal inference compared to ABMs on the ground that they rely on real-world data and do not play around with unsupported assumptions and fictional parameters.

Contrary to this view, I argue in this subsection that observational and experimental methods, too, require data that are in fact often missing, which, similarly to ABMs, implies that crucial assumptions on which these methods are built cannot practically be tested. This ultimately means that observational and experimental methods, too, cannot produce persuasive evidence of causality simply by relying on empirical data *alone*.

Below I describe a number of practical limitations arising from data availability, as typically acknowledged by (some of) the (finest) proponents of observational and experimental methods themselves. I do this systematically for RCTs, IVs, and DAGs.

5.2.1 RCTs and Heterogeneity

Since RCTs are regarded as "the failsafe way to generate causal evidence" in many disciplines (Antonakis et al. 2010: 1086), let me consider this method first.

The idea behind RCTs is simple: probabilistic relevance under intervention on a population of interest is strong evidence for causality. Suppose in a test population *all possible causes* of an outcome O (say, new cases of SARS-CoV-2 infection) are held fixed, with the exception of the putative cause, a treatment T (say, mandatory mask wearing), and it is observed that $Pr(O|T) < Pr(O)$. Then, T, i.e. mask wearing, is regarded as the cause of O in that population (for an RCT on this phenomenon, see Bundgaard et al. 2021; for a quasi-natural experimental design, see Mitze et al. 2020).

Obviously, often we have no knowledge of *all possible causes* that might confound the result. RCTs sidestep the problem. Units of analysis are randomly assigned to the treatment so that any possible systematic association between *unobserved factors* and the effect of interest is broken down (see Gangl 2010: 26). Thus, randomization allows one to treat a given population as a black box and estimate supposedly reliable mean responses in the presence of heterogeneous units of analysis.

As penetratingly noted by Cartwright (2007a: 19), however, RCTs are only apparently a self-sufficient causal machinery. Even when one is only interested in an average treatment effect, the generalizability of its statistical estimate to some target population ultimately rests on "auxiliary assumptions" that "are very demanding, demanding of information that is not supplied by the RCT and that is hard to come by". Among these assumptions, Cartwright (2007a: 16–17) noted, those that concern the way the effect of the putative cause T on O varies across subgroups of the target population are crucially important—and so are, one may add, assumptions concerning the supposed lack of temporal heterogeneity of the effect within units of analysis (on the temporal stability assumption, see already Holland 1986: 948; more recently, see Sampson et al. 2013: 13, 18–19). These assumptions are necessary because, in order to scrutinize directly the population's causal structure, one should be in a position to assess individual-level effects, but this is impossible because the treatment and its absence cannot be observed on the same individual at the same time—i.e. a fact known as the "fundamental problem of causal inference" (see Holland 1986: 947).

Not surprisingly, turn-arounds for tackling individual-level heterogeneity have been devised. For instance, the "set identification" approach, which Manski (2003) originally formulated with observational data in mind but lately extended to the counterfactual analysis of treatment responses within an experimental setting (see Manski 2007: chs. 7, 10; 2013: 63–76), renounces to point estimates of average effects and only identifies (non-parametric) regions in which these estimates may be reasonably found. This result can be reached by formulating weaker assumptions on how different subgroups of a given population react to the treatment (for a clear description of these assumptions, see Morgan and Winship 2014: 425–7). At the same time, Manski admitted that the "credibility" of these assumptions is a "subjective matter" (Manski 2003: 1). More generally, with the wisdom that typically characterizes method-creators, he claims: "An assumption is credible to the degree that someone thinks it so" (Manski 2003: 48).

Thus, the widespread view that RCTs provide the "gold standard" for causal inference, to be approximated in non-experimental contexts, too (see Angrist and Pischke 2010), essentially because RCTs would establish conclusions in virtue of assumptions that do not require any domain-specific knowledge (a claim especially overtly made by Duo et al. 2008) seems inaccurate. Causal inference in RCTs in fact requires empirical information that is often missing, thus ultimately requiring assumptions that limit the external validity of the inferences that can be made—a problem that is especially relevant when the putative causal effect is exploited to motivate policy intervention (see Cartwright and Hardie 2013). For *in practice* untestable assumptions, only theoretical reasoning, substantive knowledge and/or expert judgments can be used to make the assumptions credible. Similarly to ABMs, causal inferences, in the case of RCTs, on "horizontal" real-world connections (in Chapter's 1 terminology) can only be made on the basis of a complex mix of limited data and arguments.

5.2.2 IVs and the "Relevance" Condition

The same problem reappears with observational methods that try to recreate experimental conditions in non-experimental settings. In this case, the way the units of analysis are assigned to the putative causal variable of interest (the treatment) is not controlled by the researcher, which implies that unconfoundness cannot be claimed by construction (Imbens and Rubin 2015: ch. 3). When typical conditioning strategies, such as matching or regression (see Morgan and Winship 2014: chs. 5–7) cannot be used, because either the assignment mechanism is unknown or, if it is known, data are missing on relevant controlling covariates, one possibility is to rely on IVs (for a detailed overview, see Bollen 2012).

As claimed by Angrist and Krueger (2001: 73, emphasis added), "[t]he instrumental variables methods allow us to estimate the coefficient of interest consistently and free from asymptotic bias from omitted variables, *without actually having data* on the omitted variables or *even knowing what they are*". Thus, in principle, by exploiting the existence of variables, whose properties resemble the properties of intervention variables, the technique promises to help draw causal inferences without relying on extensive background knowledge about the causal structure.

More precisely, given a putative causal relation between T (say, economic aids) and O (say, economic growth), the IV approach purports to identify the causal effect of T on O, based on the existence of an IV I (say, an earthquake disrupting the aids as a function of their distance from its epicenter) that is (i) highly correlated with the putative cause T, and (ii) uncorrelated with the putative effect O given T. The property (i) is typically called the "relevance" condition, whereas property (ii) is referred to as the "exogeneity" or "exclusion" restriction (see, respectively, Stock and Watson 2010: 333; Gangl 2013: 381). The crucial point is that, to obtain reliable estimates, both conditions must be satisfied.

Now, to what extent do empirical data *alone* allow the researcher to judge whether or not this is the case, thus ensuring that it is indeed possible to draw inferences "(...) *without actually having data* on the omitted variables or *even knowing what they are*", as claimed by Angrist and Krueger's above-mentioned statement? Let me consider first the "relevance" condition—I will discuss the "exclusion" restriction in Section 5.3.2 when I will treat assumptions that are unstable in principle rather than being untestable in practice because of lack of data.

As noted by Bollen (2012: 59), the relevance condition has started to receive serious attention only later than the exclusion restriction (for a recent summary, see Hernán and Robins 2020: 204–5). Although apparently innocuous, the relevance condition is a crucial assumption indeed. In an important article, Bound et al. (1995: 445) re-examined Angrist and Krueger's (1991) seminal applications of IVs in economics and showed that, with a finite sample, IV estimates are "(...) biased in the direction of the expectation of the OLS estimator". The point is that the size of this bias increases as a function of the strength of the correlation between the IV(s) and the endogenous variable(s). When the instrument is "weak", i.e. only marginally correlated with the putative cause(s) of interest, "even enormous samples do not eliminate the possibility of quantitatively important finite-sample biases" (Bound et al. 1995: 446). This point would be later easily admitted by Angrist and Krueger (2001: 79) themselves. In addition, Bound et al. (1995: 444) show that, when an instrument is weak, small violations to the exclusion restriction are amplified, thus leading to even larger biases.

But how do we know whether an instrument is weak? Given how consequential a violation of the relevance condition can be, it is not surprising that a large literature has emerged on strategies to empirically assess the strength of an instrument (for an overview, see Stock et al. 2002: §4; Stock and Yogo 2005) as well as on methods that are supposed to be robust against weak instruments, at least in large samples (for an overview, see Stock et al. 2002: §5–§6). Within this literature, Steiger and Stock (1997) played an important role in diffusing the following "rule-of-thumb", which seems now ubiquitous in econometric textbooks: a first-stage partial-F test of less than 10 indicates the presence of weak instruments (cf. Stock and Watson 2010: 350).

However, careful reading of Stock et al. (2002: 521–2) suggests that weak instruments can be conceptualized in two different ways—one based on relative bias and the other on size—such that the statistical test leads to different threshold values, which in turn are a function of the number of instruments considered. Moreover, when Stock and Yogo (2005) discussed their rule of thumb in comparison with other possible test procedures, each providing its own rejection thresholds (see, in particular, Stock and Yogo 2005: tables 5.1–5.4), they qualified their justification for the 10-value threshold as "not unreasonable" and acknowledged that "(...) when the number of instruments is moderate or large, the critical value is much larger and the rule of thumb does not provide substantial assurance that the size distortion is controlled" (*ibid.*: 103).

In the light of this, it seems fair to say that, although the relevance condition is in principle empirically testable, available procedures based on the data that can be typically accessed cannot have the last word. Bollen (2012: 59) noted that "diagnostics and tests for weak [instrumental variables] continue to evolve"; Stock (2001: 7581) remarked that "(...) no single preferred way to handle weak instruments has yet emerged (...)"; and Bound et al. (1995: 449) explicitly recommended that tests for weak instruments be used only as "rough guides". More generally, Angrist and Krueger (2001: 76) explicitly rejected "one of the most mechanical and naïve, yet common, approaches to the choice of instruments", namely one that "uses atheoretical and hard-to-assess assumptions about dynamic relationships to construct instruments from lagged variables in time series or panel data".

After all, that data *alone* cannot provide a conclusive argument in favor of the strength of an instrument is nicely illustrated by Angrist and Krueger's (1991) pioneering use of the quarter of birth as an instrument for estimating the causal effect of schooling on earnings. The empirical correlation between the instruments and the endogenous variables, i.e. schooling, was very low (R2 between 0.001 and 0.002), thus providing limited support for the relevance condition. Actually, the justification for the instrument's choice essentially came from the argument and the independent empirical evidence that the authors were able to offer to the reader. And it is arguably thanks to this reasoning that the article has become a classic in the IV literature, even if the instrument was proven to be sub-optimal on pure statistical grounds (Bound et al. 1995).

Thus, similarly to ABMs, causal inference based on IV methods relies on assumptions like the relevance condition of the chosen instrument that, while conditioning the credibility of the inference, cannot be fully justified through empirical data *alone* but requires independent theoretical reasoning, substantive knowledge, or expert judgments. As noted by Angrist and Krueger (2001: 73), "(...) good instruments often come from detailed knowledge of the economic mechanism and institutions determining the regressor of interest".

5.2.3 DAGs, Causal Discovery Algorithms, and Graph Indistinguishability

In a recent overview chapter on causal inference and estimation, Richard Breen (2022) rightly noted: "A distinction exists between, on the one hand, estimation methods that seek to minimize the role of theory in establishing a causal relationship by trying to mimic an RCT as closely as possible, and, on the other, approaches that are explicitly based on, and whose causal claims are conditional on, theory. The distinction is not absolute, however, because all attempts to infer and estimate causal quantities from observational data depend on assumptions to some degree."

The research stream that exploits DAGs as support for computerized causal discovery algorithms—see, in particular, Spirtes et al. (2000), Pearl (2009), Korb and Nicholson (2011), and, for a recent overview, Glymour et al. (2019)—is clearly close to the atheoretical extreme of Breen's continuum, and, in my view, constitutes another especially meaningful example of the actual impossibility for causal inference based on observational methods to exclusively rely on empirical data.

In general, a causal graph is a (non-parametric) mathematical object made of vertices and (missing) edges, where the former represent variables and the latter represent (the absence of) connections among these variables. When a directed edge from vertex A to vertex B exists, A is said to be a parent of B, and B is said to be a child (or descendant) of A. So-called DAGs are a special class of graphs, such that all edges are directed, and there are no directed cyclic paths (Pearl 2009: ch. 2). Minimally, two fundamental conditions must hold for the edges of a DAG to be causally interpretable (Spirtes et al. 2000: 11, 13; see also Spirtes 2010: 1651, 1654): (i) each variable in the graph must be independent from its non-descendants given its parents ("Markov condition") and (ii) all the (conditional) independences among variables are implied by the Markov condition applied to the graph under study ("faithfulness").

In the 1990s, a research program at the intersection of philosophy and computer science proposed causal graphs as a tool for automating the discovery of causal structures when background knowledge is scarce and/or uncertain. Spirtes (2010: 1648) provides a clear statement of the motivation for this research program that it is worth quoting *in extenso*:

> In new domains such as climate research (where satellite data now provide daily quantities of data unthinkable a few decades ago), fMRI brain imaging, and microarray measurements of gene expression, the number of variables can range into the tens of thousands, and there is often limited background knowledge to reduce the space of alternative causal hypotheses. (...) In such domains, non-automated causal discovery techniques from sample data, or sample data together with a limited number of experiments, appear to be hopeless, while the availability of computers with increased processing power and storage capacity allow for the practical implementation of computationally intensive automated search algorithms over large search spaces.

Despite the variety of algorithms proposed (for a detailed description, see Spirtes et al. 2000: ch. 5; for a software perspective overview, see Kalisch et al. 2012), the common logic of the approach is the following: given the distribution over observed variables, the algorithm iteratively deletes and orients edges as a result of statistical tests for conditional independence (constraint-based search algorithms) or as a result of changes in given model selection statistics, typically the Bayesian information criterion (score-based search algorithms). On these technical bases, the proponents of algorithmic search methods from observational data contend: "given our assumption, with an oracle that can correctly answer questions about conditional independencies and dependencies in a population, the outputs of our algorithms are correct" (Spirtes et al. 1997: 561).

Despite these claims, the muscled exchange between Freedman and Humphreys (1996, 1999) and Spirtes et al. (1997) suggests that DAG-based, automatized causal discovery has two deep limitations. In particular, it can be argued that, first, in practice, data *alone* are not sufficient to generate reliable causal models, and, second, substantive, domain-specific, and expert judgments enter the search process at different stages.

As to data insufficiency, Freedman and Humphreys (1996: 117) highlighted that the core of causal discovery algorithms, i.e. testing for conditional independence, is fragile when working with real data: "exact conditional independence cannot be determined from any finite sample. (...) correlations of 0.000 and 0.001—at the population level—play very different roles in Spirtes, Glymour, and Scheines's theory. A sample of realistic size cannot distinguish between such correlations" (cf. Freedman and Humphreys 1999: 37). As a consequence, they argued, causal discovery algorithms tend to output different causal models as a function of the significance level and assumptions behind the chosen statistical test (Freedman and Humphreys 1999: 42). Gelman (2011: 961) formulated the same objections when he noted that "the difficulty is that, in social sciences, there are no true zeros (...) anything that plausibly could have an effect will not have an effect that is exactly zero".

Proponents of computerized causal discovery algorithms recognized this limitation. As noted by Kalisch et al. (2012: 2), "(...) in general one cannot estimate a unique DAG from observational data, not even with an infinite amount of data, since several DAGs can describe the same conditional independence information". More generally, Spirtes et al. (1997: 561) conceded that "given the problem of sampling error, no algorithm whose output is a function of the sample could guarantee success" (see also Spirtes et al. 2000, 87–90, 296, 350–1; Spirtes 2010: 1656; Spirtes et al. 2000: ch. 4, for a discussion of different forms of graph indistinguishability). As a partial solution, Spirtes et al. (1997: 562) suggested that simulation can be used to assess the robustness of the algorithms, and report on sensitivity tests that indicate that for samples larger than 2000 observations, when variables have no more than two or three parents, errors concerning presence/absence of connections among variables are rare whereas those concerning causal direction are more frequent.

Moving on to the second limitation, namely the role of background knowledge, critics of causal discovery algorithms noted that the method's followers tend to underestimate the role of theoretical inputs (see Freedman and Humphreys 1999: 29, 41), although these inputs are in fact necessary to overcome data limitations and guarantee the method's reliability. As Freedman and Humphreys (1999: 40) put it, "causal discovery algorithms succeed when they are prevented from making mistakes". On this point, Spirtes et al. (2000: 93) *de facto* allowed the algorithms to incorporate prior knowledge about the existence or nonexistence of certain edges in the graph, the orientation of some of the edges, or the time order of the variables. As Spirtes (2010: 1654, 1655) overtly admitted, background knowledge is ultimately necessary to select the most plausible causal graph among the set of equivalent graphs the algorithm generates.

Both technical and theoretical limitations of DAG-based causal search algorithms are well summarized by Quintana (2020) in the conclusion of a recent, and one of the rare, applications of this approach to causal inference, in social sciences:

> Even if I included a wide range of variables measuring relevant psychological and contextual factors, the presence of unmeasured confounders remains a likely possibility. Confounding variables can generate erroneous graphs, for example, by including spurious direct connections (Spirtes et al. 2000). Additional errors in the

estimated graph might be generated by lack of statistical power, violations of faith-fulness and modularity assumptions, feedback loops, selection bias, measurement error, and inadequate time intervals. Given that—to a greater or lesser extent—these assumptions could have certainly been violated in the present analysis, the estimated graphs should not be considered representations of the true causal struc-tures. It is worth mentioning, then, that causal search algorithms do not "solve" the problem of causal inference and that causal inference from observational data remains, as several authors have noted (e.g. Dawid 2008; Greenland 2010), a largely speculative and difficult to validate exercise.

Thus, although popularized as a fully inductive, data-driven approach to causal inference from observational data, DAG-based causal discovery algorithms, not unlike ABMs, can-not rely on data *alone*, and in fact require independent theoretical reasoning, substantive knowledge, and expert judgments to single out persuasive causal models.

5.3   *In Principle* Untestable Assumptions

As suggested by the exchange between Marshall and Galea (2015) and Diez Roux (2015) that I commented on in the chapter's introduction, advocates of observational and experi-mental methods tend to believe that, differently from the assumptions used in ABMs, those used in observational methods are testable with data. Contrary to this view, I have just shown that some crucial assumptions that are necessary to causally interpret the results produced by RCTs, IVs, and DAGs cannot in fact be verified empirically because of practical data limitations. Thus, as is the case with ABMs, theoretical and/or empirical knowledge external to the method is ultimately required to make the *horizontal* evidence produced by the method conclusive.

I want now to make an even stronger point. In particular, I shall argue that some of the assumptions on which observational and experimental methods rely to establish (hori-zontal) causal claims are in fact *in principle* untestable. By this I mean that the only condi-tion that would make some assumptions verifiable is a perfect and complete knowledge of the myriad of events affecting a given phenomenon. Given that this condition is obviously never met, the untestable character of the assumption does not derive here from the amount of data available—as for *in practice* untestable assumptions discussed in Section 5.2—but from an *a priori* methodological impossibility. In other words, data will never tell us, by construction, if the assumption is tenable or not.

Thus, I shall argue, irrespective of data availability, causal inference by observational and experimental methods will always have leaps, just as causal inference from ABMs, due to the unavoidable uncertainty as regards the truth of some of the assumptions on which the methods rely.

5.3.1   RCTs and "Stable Unit Treatment Value Assumption" (SUTVA)

Again, let me first consider RCTs. As mentioned, this method is considered the gold standard for causal inference because unit randomization is believed to solve the problem of confounding by design—a fundamental belief itself that also begins to be critically scrutinized (see Deaton and Cartwright 2018: §2).

In Section 5.2.1, with respect to the potential heterogeneity of causal effects within the population under scrutiny, building on Cartwright's (2007a) critique I have highlighted that RCTs are in fact only apparently self-sufficient causal machinery. The reason is that, in order to generalize the results to a target population, assumptions on the causal structure of the population under examination, which are difficult to test empirically, are required (for a similar claim as to RCTs in development economics, see Deaton 2010). In this subsection, I consider a more general and deeper assumption of the experimental approach, the so-called "stable unit treatment value assumption" (SUTVA) (for other labels, see Morgan and Winship 2015: 48), which, as noted by Gangl (2010: 38), is "much more restrictive and problematic than is commonly recognized in the discipline".

According to Imbens and Rubin's (2015: 10) recent formulation of an idea originally labeled and made explicit by Rubin (1980: 591), the SUTVA requires that "the potential outcomes for any unit do not vary with the treatments assigned to other units, and, for each unit, there are no different forms or versions of each treatment level, which lead to different potential outcomes". Imbens and Rubin insisted on the fact that the assumption contains two components, the first clearly referring to potential interferences among units and the second concerning potentially unnoticed "hidden variations of treatments".

The point obviously is that, when experimenting with humans, both requirements can easily be violated. As to the supposed absence of interferences, violations typically obtain when direct or indirect social interactions are at work; as to the supposed absence of unobserved treatment heterogeneity, violations can arise from variations in the quality of the treatment, in the quality of those who administer it, or in subtle behaviors endorsed by the units themselves (Imbens and Rubin 2015: 10–12). The SUTVA violations are consequential because, in the presence of interferences and/or hidden treatment heterogeneity, one may attribute to the treatment a causal impact that in fact arises from other unobserved and subtle processes.

An example can clarify the point. In his critical analysis of studies assessing the causal impact of residential relocations within the "Movement to Opportunity" (MTO) in-the-field experimental program, Sobel (2006: 1401) noted that "because the ITT [intent to treat] and TOT [treatment on the treated] are defined assuming no interference, and this assumption is not reasonable, it is not clear what parameters MTO researchers are estimating or what policies their analyses might support". Sobel (2006: 1403–5) formally showed that, when social interactions are taken into account, the average treatment effect in fact amounts to the difference between the average effect of the treatment and "spillover effects on the untreated". According to Sobel (2006: 1399), one may expect this situation to be very common in the social world because "interference is the norm" (see also Gangl 2010: 38). That is why Hong and Raudenbush (2013: 337) claims that "(...) SUTVA, while arguably plausible in the classic clinical trial, appears highly implausible and often consequential in social settings" (for a similar observation, see Makovi and Winship 2021; Breen 2022).

When violations of the SUTVA are foreseeable, an RCT can be devised in such a way as to prevent these violations. For instance, a common strategy to achieve this goal is to design the experiment so that the units of analysis are assigned to clusters within which the assumption of absence of social interactions and treatment homogeneity can be more easily defended (for an overview of this type of design, see Hong and Raudenbush 2013). But, and this is my point, these designs, too, rely on assumptions that may be difficult to test empirically and, as noted by Imbens and Rubin (2015: 11), to some extent some "more distant" versions of the SUTVA must be posited (for a review of two-stage randomization, see Halloran and Hudgens 2016; I will discuss this specific solution in Chapter 6, Section 6.5).

Then, on what grounds is the plausibility of this or that form of the SUTVA ultimately established? Given the nature of the potential biasing processes at issue, fueled by diffuse social interdependences and individual-level reactions that are difficult to trace, empirical data can hardly have the last word. There will always be some uncertainty about whether violations of the SUTVA do occur or not. That is why Imbens and Rubin (2015: 10, emphasis added) themselves claimed:

> SUTVA exemplifies assumptions that rely on *external, substantive, information* to rule out the existence of a causal effect of a particular treatment relative to an alternative. (...) these assumptions, and other restrictions discussed later, *are not informed by observations*—they are assumptions. That is, they rely on *previously acquired knowledge of the subject* matter for their justification. Causal inference is generally impossible without such assumptions, and thus it is critical to be explicit about their content and their justifications.

Thus, similarly to ABMs, even RCTs, the method that is regarded as the gold standard for causal inference, in fact rely on assumptions like the SUTVA, which cannot be empirically tested by construction, so to speak, because the verification of this type of assumption would require a full knowledge of the inner functioning of the social world. For these *in principle* untestable assumptions, theoretical reasoning, substantive knowledge and expert judgments are *a fortiori* necessary to reduce the uncertainty on the occurrence of violations. Causal inference in design-based identification strategies, too, is not only a matter of data: limited empirical information, plus arguments, must be creatively combined.

5.3.2 IVs and the "Exclusion" Condition

IVs provide a further clear illustration of this challenge, which observational methods, too, have to face. When randomization is not possible, this method promises to sidestep the problem of the practical impossibility of controlling for all potential confounders by exploiting a variation that is "external" to the system (of variables) under scrutiny. Causally interpretable estimates are conditional, however, on two assumptions, which are, to recall, the "relevance" condition (the instrument I must be highly correlated with the putative cause T) and the "exogeneity", or "exclusion", condition (I must be uncorrelated with the putative effect O given T). In Section 5.2.2, I discussed the practical difficulties of empirically ascertaining whether or not the former assumption is met. Here I want to stress that the obstacles in verifying the latter are even more severe. In fact, it can be argued that the "exogeneity" assumption is an *in principle* untestable assumption. As noted by Deaton (2010: 431), the problem with this assumption is that "exogeneity is an identifying assumption that must be made prior to analysis of the data, empirical tests cannot settle the question". Essentially, instrument exogeneity requires that all potential pathways going from I (i.e. the instrument) to O (i.e. the potential outcome) are controlled for. This is a condition that, by construction, cannot be verified relying on the data under scrutiny, and that, since it is always possible that a confounder is not taken into account, is in principle unverifiable. That is why Gangl (2013: 381, emphasis added), more overtly than others, admitted that "(...) it is important to realize that the exclusion restriction is an assumption that *is not testable in principle*" (see also Hernán and Robins 2020: 194, 205).

Morgan and Winship (2015: 301–2) provided a clear explanation of why this is the case. They show that the still common strategy of conditioning on T (i.e. the treatment) and checking whether I and O are independent cannot provide an empirical test of the exogeneity condition. The reason is that an association between I and O will exist not only when the instrument fails to be exogenous but also when the instrument is valid and yet T is a "collider", i.e. a variable correlated with both I and unobserved factors highly correlated with the putative effect (on the notion of collider, see, in sociology, Elwert and Winship 2014). As to models including several instruments, if their number is higher than those of putative causes included in the model, the so-called "over-identification" test can be performed (see, for example, Stock and Watson 2010: 353–4). However, as noted by Bollen (2012: 56), this statistical test can only suggest that some instruments are correlated with the error term; it cannot clearly say *which of them* is violating the exogeneity condition.

As a consequence, "assessing whether the instruments are exogenous *necessarily* requires making an expert judgment based on personal knowledge of the application" (Stock and Watson 2010: 353, emphasis in original). Even more fundamentally, as remarked by Deaton (2010: 432), it requires "thinking about how and why things work". This conclusion was also reached by re-evaluating studies that exploited naturally occurring events as instruments (see, in particular, the statement on "plausible stories" by Rosenzweigh and Wolpin 2000: 830).

It is important to note that the exogeneity condition must also hold for IV estimators that are devised to account for potential treatment effect heterogeneity in the population under scrutiny, i.e. the so-called LATE ("local average treatment effect") estimators (for a clear introduction, see Morgan and Winship 2015: 305–15).

This is another interesting case because proper estimation of LATE requires an additional assumption that must be made prior to the data analysis, namely the "monotonicity" assumption according to which the instrument is supposed to affect all individuals either positively (if they are "compliers") or negatively (if they are "defiers") but not both at the same time. The assumption is *in principle* empirically untestable, however, because, as clarified by Gangl (2010: 37), "LATE is based on changes in the expected exposure to treatment, not on actually observed changes of treatment status" (see also Hernán and Robins 2020: 200–3). This leads some to stress again the role of theoretical reasoning to complement econometric assumptions so as to deal with heterogeneous causal effects. In Deaton's (2010: 430) words: "heterogeneity is not a technical problem calling for an econometric solution but a reflection of the fact that we have not started on our proper business, which is trying to understand what is going on" (see also Angrist and Krueger 2001: 78).

Thus, to sum up, fundamental assumptions—like the "exclusion" restriction and the "monotonicity" assumption—required by IVs to generate reliable, and causally interpretable, estimates are *in principle* untestable assumptions, i.e. cannot be tested by recourse to data. In the recent words of Breen (2022):

> Even in the case of instrumental variables the exclusion restriction has to be justified; a notable feature of many IV papers is that details of the estimation take up far less space than explanations of why the exclusion restriction is warranted. Identification can never be proved definitively when we have observational data; it is always conditional on the assumptions made. Ideally these assumptions would be grounded in strong theory.

Causal inference that relies on IVs has leaps, just as does causal inference that relies on ABMs, due to the unavoidable uncertainty as regards the truth of some of the assumptions on which the method relies. Theoretical reasoning, substantive knowledge, and/or expert judgments are necessary to reduce such uncertainty.

5.3.3 DAGs and Strategies for Causal Identification

Compared to methodological developments in computer science that I have discussed in Section 5.2.3, where DAGs are used as a means for inductively discovering causal relations from data through automatized algorithms and face practical difficulties imposed by data scarcity, in sociology DAGs are spreading in association with the counterfactual view of causality (for a clear introduction, see Morgan and Winship 2015: ch. 3), and emphasis is put on using DAGs as a deductive tool that allows one to clearly express sets of substantive and domain-specific hypotheses, and to establish explicitly under which conditions the associated causal connections can be identified (cf. Elwert 2013: 246–7).

This use of DAGs has the virtue of making even more visible the presence of assumptions, which are necessary for causal inference and yet untestable *in principle*, in particular when experimental designs are not realizable. Next I discuss this point with respect to two fundamental strategies for causal identification known as the "backdoor" (Section 5.3.3.1) and the "front door" (Section 5.3.3.2) criterion.

5.3.3.1 DAGs and the "Backdoor" Criterion

To introduce the concept of "backdoor" criterion, let us start with "the traditional way to deal with the potential problems raised by noncausal pathways", namely "conditioning" (Knight and Winship 2013: 286).

As pedagogically clarified by Morgan and Winship (2015: 128–9), this identification strategy takes the form of "balancing" or "adjusting", and is implemented in a large variety of statistical methods for causal identification from observational data, ranging from matching techniques to regression-like models (Morgan and Winship 2015: 128–9, chs. 5–7). Instead of adopting the defensive strategy of "controlling for everything", DAGs support a principled definition of the minimal set of variables that is sufficient to make conditioning effective.

The "backdoor criterion" is precisely the most fundamental of these principles (Pearl 1993: 268), according to which a conditioning set Z is sufficient if three conditions are satisfied: (i) it blocks all "backdoor" paths from the treatment T to the outcome O (i.e. the non-causal paths that start with an arrow into T), since these introduce confounding; (ii) it does not contain any collider (i.e. an endogenous variable that has two or more causes) lying on backdoor paths, since conditioning on a collider creates new conditional dependences; and (iii) it contains no descendants of T, since these cancel out some portion of the causal effect of T on O (cf. Elwert 2013: 259; Morgan and Winship 2015: 109–17, 130–9).

But is the backdoor criterion assumption-free? The first condition seems the most problematic. The application of the backdoor criterion in fact requires that the variables that are necessary to block the backdoor paths are observed. When discussing the now classical study by James Coleman and colleagues of the putative causal effect of Catholic schooling on students' learning, Morgan and Winship (2015: 122; see also 126, 130) themselves admitted this challenge: "(...) in most cases (...) the directed graph will show clearly that only a subset of the variables in [schooling] that generate confounding are observed, and the confounding that they generate cannot be eliminated by conditioning with the observed data".

Formulated in these terms, the difficulty with the backdoor criterion seems limited to a practical problem of data limitation. I contend, however, that the challenge is much deeper. Indeed the set(s) of variables this principle recommends to condition on in a particular application in order to achieve identification crucially depends on the variables included in the graph. Thus, the underlying, fundamental assumption is that *all* relevant variables are present. In other words, the causal graph must be right. In Elwert's (2013: 249) words, "when working with DAGs, the analyst (for the most part) needs to assume that the DAG captures the causal structure of everything that matters about a process". Similarly, Hernán and Robins (2020: 90) observed: "The procedure requires *a priori* knowledge of the causal DAG that includes all causes—both measured and unmeasured—shared by the treatment A and the outcome Y". This obviously is a meta-assumption that, *in principle*, cannot be tested empirically. Only external, theoretical, substantive, and domain-specific knowledge can help to reduce the uncertainty about the truth of the causal graph under examination.

5.3.3.2 DAGs and the "Front Door" Criterion

The "front door" criterion (Pearl 1995: 676) is the second fundamental DAG-based identification strategy that must be considered because it promises to identify causal effects in settings where data are missing on known confounders, thus making the backdoor criterion inapplicable (see Knight and Winship 2013: 287–8). In this sense, the front door criterion pursues the same goal as IVs (for a DAG-based representation of IVs, see Makovi and Winship 2021: fig. 7).

The idea behind the front door criterion is to condition on a set of intervening variables Z between the treatment T and the outcome O such that (i) the chain of intervening variables captures all the directed paths from T to O, (ii) there are no unblocked paths connecting T to Z, and (iii) conditioning on T blocks all backdoor paths connecting Z to O.

The first condition is referred to as "exhaustiveness"; the second and the third define "isolatability" (see Morgan and Winship 2015: 333–4). As we have seen in Chapter 1 (Section 1.3.2), according to scholars sharing the "horizontal" view of mechanisms, a chain of intervening variables that is exhaustive and isolated constitutes a "mechanism". That is why they sometimes also speak of "identification by mechanisms" in relation to the front door criterion (see, for instance, Knight and Winship 2013).

But, again, on what grounds can the exhaustiveness and isolatability assumptions be justified? The DAG under study cannot by itself say anything about the truth of these two assumptions because, as was the case with the backdoor criterion, the DAG must *a priori* be supposed to be valid for the front door criterion to deliver any reliable conclusion (see Elwert 2013: 261). Can then empirical data external to the DAG under scrutiny help? In my view, they cannot. The assumption that the chain of intervening variables describes all causal pathways from T to O (i.e. "exhaustiveness") is so demanding that no empirical data can be realistically imagined to support them. Similarly, given the complexity of social phenomena, it seems highly implausible that empirical data *alone* can exclude that some unobserved factors open backdoor paths to the set of intervening variables and/or to the outcome of interest (i.e. "isolatability").

That is probably why Morgan and Winship (2015: 337) ultimately remarked, with reference to a weak form of exhaustiveness, that "to assert this assumption, one typically needs to have a very specific theoretical model of the complete identifying mechanism, even though part of it remain unobserved", and, with reference to isolatability, Knight and Winship (2013: 293) noted that "more generally, the point is that it is necessary to specify the mechanisms involved in a causal process in sufficient detail (or in sufficient depth) so that (some of) the mediating variables involved are isolatable". But it seems fair to say that both operations are intrinsically theoretical.

Thus, similarly to causal inference from ABMs, causal inference based on observational methods has leaps. DAGs clearly show that these methods require assumptions that are *in principle* untestable with data, hence requiring theoretical, substantive, and domain-specific knowledge as ultimate elements to convincingly show that the assumptions on which the identification strategy relies are satisfied. Again, causal inference is matter of limited data, plus arguments.

5.4   Are ABMs, Experimental, and Observational Methods Fundamentally Similar?

In the previous two sections, I considered experimental (namely, RCTs) and observational (namely, IVs and DAGs) methods illustrating typical identification strategies of causal effects, and I have emphasized that these methods cannot support causal claims on robust or counterfactual dependences relying *only* on empirical data. In particular, I discussed crucial assumptions that are necessary to causally interpret the results of these methods and showed that these assumptions are either *in practice* untestable (i.e. they cannot be verified empirically because of practical data limitations) or *in principle* untestable (i.e. the test of these assumptions would require a perfect and complete knowledge of the myriad of events affecting the phenomenon under examination, which is not accessible by construction). As a consequence, I have argued, background knowledge, which is external to the method the researcher is using, must be mobilized to complement what data *alone* can tell us.

As shown by several quotes that I commented on in the previous sections, the most expert experimentalists as well as users of multivariate statistics are ready to acknowledge this fact. However, I suspect that they would be more reluctant to accept the implication that, in my view, should be drawn from it: that experimental and observational methods are fundamentally similar to ABMs with respect to their capacity to sustain persuasive causal claims because, in both cases, causal inference (along horizontal and vertical lines respectively) has leaps, thus making theoretical, substantive, and domain-specific knowledge necessary to make the inference credible when evidence is insufficient or by-construction inexistent.

Diez Roux's (2015: 101) claim that I quoted in the introduction to this chapter suggests that experimentalists and observational method users would be likely to argue that the similarity in fact is only apparent in that (i) the "formal" nature of the assumptions on which experimental and observational methods rely is fundamentally different from that of ABMs; (ii) the "materiality" of the systems studied by experimental and observational methods make them ultimately "superior"; and (iii) the reliability tests these methods can perform are in any case more solid.

Thus, to conclude this chapter, I discuss in the next three subsections each of these putative fundamental differences between ABMs and experimental/observational methods, and explain why, in my view, it is unjustified to claim that as a matter of principle these methods still are more reliable tools for causal inference despite the fact that, similarly to ABMs, they cannot entirely rely on empirical data to establish claims on robust and counterfactual dependences.

5.4.1 Objection 1: ABM Lacks "Formal" Assumptions

By introducing causal graphs (see Section 5.2.3), I explained that the two conditions (namely, the "Markov" and "faithfulness" conditions) must hold for the edges of a DAG to be causally interpretable. It is now instructive to consider how Glymour and Greenland (2008: 191) qualified these two assumptions:

> The rules and assumptions just discussed should be clearly distinguished from the content-specific causal assumptions encoded in a diagram, which relate to the substantive question at hand. These rules serve only to link the assumed causal structure (which is ideally based on sound and complete contextual information) to the associations that we observe. In this fashion, they allow testing of those assumptions and estimation of the effects implied by the graph.

Glymour and Greenland's remark is important because it clearly illustrates the general way the proponents of experimental and observational methods for causal inference interpret the assumptions—such as the SUTVA, exclusion restriction, Markov condition, etc.—or the requirements—such as the "backdoor" and "front-door" criteria—on which these methods are based. According to experimentalists and statistically minded scholars, these assumptions and requirements are "formal" in the sense that they do not purport to depict the mechanism of interest but only act as bridges between the chains of connections to be identified and the available data. Formal assumptions and requirements within experimental and observational models have the role to make the inference valid.

In contrast, an ABM is "a virtual world (based on prior knowledge or intuition)" (Diez Roux 2015: 101) where everything is dependent on "substantive" assumptions, i.e. theoretical, domain-specific hypotheses on the particular mechanism of interest. A recent comparison of DAG-informed regression modeling and ABMs clearly reiterated the point when the authors noted that the former approach is "backed by formal mathematics of graphical model theory" and "assumptions underlying each model are transparent" whereas ABMs "lack consensus about fundamental assumptions" (see Arnold et al. 2019: 251). Thus, although ABMs may be internally valid models, the method itself lacks "formal" assumptions by which to establish whether the models are externally valid, too. That is why ABMs must exploit knowledge that comes, and cannot but come, from outside the method, namely data produced through experimental and/or observational methods.

The view that an ABM is entirely made of "substantive" assumptions seems to me simply wrong from a technical point of view. When I introduced the concept of empirical calibration of an ABM (see Chapter 4, Section 4.3), I noted that the distinction between "substantive" and "formal" assumptions within an ABM must be framed in terms of the role a given assumption plays within the model rather than in terms of the assumption's content. From this perspective, within an ABM, any assumption for which data are insufficient, so that the model's components implementing that assumption cannot be (entirely) grounded on real-world information, is to be considered "formal" in the sense that this assumption limits the validity of the causal claims one can make as to the connection between the empirically calibrated parts of the model and the model's outputs at a higher level of abstraction. Thus, "formal" assumptions within experimental and observational methods are different from "formal" assumptions within an ABM because, in the former case, those assumptions are generic (meaning, they are the same for any substantive chains of variables to be identified and estimated) whereas, in the latter case, "formal" assumptions may vary from one model to another. However, in both types of methods, "formal" assumptions play exactly the same role. Depending on how well and extensively they are justified, and their consequences assessed, these assumptions make more or less credible the causal interpretation of the connections described by experimental/observational methods and generated from the bottom-up by an ABM.

Within ABMs, it is clear that a "formal" assumption is untestable (in practice or by construction). We have seen that how consequential "formal" assumptions are within an ABM can be assessed by systematic application of several reliability tools (see Chapter 4, Sections 4.3.1–4.3.4). What about experimental and observational methods? What sort of justification, if any, do the formal assumptions of observational methods require and actually receive?

In order to reliably connect models and data, the assumptions of observational methods have to be true. Furthermore, since they are not substantive assumptions, which lend themselves to testing (by relying on formal assumptions), but formal assumptions, which by their very nature have to be assumed for testing to be possible, they have to be, in a sense, *a priori* true. As to the SUTVA, for instance, Rubin (1986: 961) explicitly noted "(...)

we are not ready to estimate, test, or even logically discuss causal effects until units, treatments, and outcomes have been defined in such a way that SUTVA is plausible". But how justified is the view that this type of formal assumption is *a priori* true? The above discussion of *in practice* and *in principle* untestable assumptions within experimental and observational methods demonstrated that it is by no means obvious that the formal assumptions behind these methods can be easily regarded as universally valid.

For instance, to come back to one of the assumptions cited by Glymour and Greenland (2008), the faithfulness assumption is violated if positive and negative effects along different pathways cancel each other out (for a well-known example, see Hesslow 1976). Defenders of the faithfulness assumption argue that any parametrization violating faithfulness has Lebesgue measure zero[2] (Spirtes et al. 2000: 66). However, whether this provides a conclusive argument in favor of the assumption is questioned, too. Cartwright (2007b: 68), for example, observed:

> it is unlikely that any causal system to which we consider applying our probabilistic methods will involve genuine causes that are not prima facie causes as well. But this conclusion would follow only if there were some plausible way to connect a Lebesgue measure over a space of ordered n-tuples of real numbers with the way in which parameters are chosen or arise naturally for the causal systems that we will be studying. I have never seen such a connection being proposed; that is I think because there is no plausible story to be told.

An analogous point applies to the other assumption cited by Glymour and Greenland (2008), namely the Markov condition, which is known to fail in the physical realm, where non-deterministic common causes may fail to screen off their effects. In the biological and social realms, in any case where multiple effects of a probabilistic common cause always co-occur, it will be impossible to recover the correct causal structure by relying on the Markov condition (for a more extensive discussion, see Cartwright 1999; see also Salmon 1984: 168–9). As noticed again by Cartwright (2007b: 107):

> viruses often produce all of a set of symptoms or none; we raise the interest rate to encourage savings, and a drop in the rate of consumption results as well; we offer incentives to single mothers to take jobs and the pressure on nursery places rises; and so forth.

To avoid misunderstanding, let me clarify that these remarks are not intended to deny that these and other assumptions can be reasonable in certain cases. My point is rather that there is no *a priori* guarantee that they are *necessarily* true, or even approximately true, of the causal structures to which they are meant to apply. In this sense, the argument that the truth of ABM's formal (and substantive) assumptions is context-dependent, whereas the truth of experimental and observational methods' formal assumptions is not, is not convincing.

---

[2] A Lebesgue measure is a measure of subsets of n-dimensional Euclidean spaces, such as probability spaces over continuous random variables.

Moreover, even accepting for a moment the claim that it is less clear within an ABM compared to experimental and observational methods "under what circumstances causal effect estimates are valid" (Arnold et al. 2019: 251) because the former lacks formal assumptions that are generic (i.e. invariant across specific models), one may argue that there are good reasons to reverse the qualitative hierarchy between formal and substantive assumptions. As forcefully argued by prominent statisticians like Freedman (2009, 2010), formal assumptions, applicable across contexts, never grant by themselves the credibility of the inference from an observational method. This point is also echoed in philosophy by Cartwright (2007b: 68), who (in relation to the faithfulness assumption) stressed that: "it is not appropriate to offer the authority of formalism over serious consideration of what are the best assumptions to make about the structure at hand". I take these remarks to show that even seemingly formal assumptions carry a substantive weight and require justifications that are often context- and domain-specific if they are to reliably connect models to data.

Thus, it seems fair to conclude that the distinction between formal and substantive assumptions is a matter of degree, and that the potential generic nature of the former does not constitute a guarantee in itself of their *a priori* plausibility. There are no differences, based on the nature of the assumptions in ABMs and experimental/observational methods, which make the former in principle inferior to the latter for establishing causal claims (on vertical and horizontal mechanisms, respectively). In both cases, ultimately only a complex combination of always limited data and theoretical reasoning can secure persuasive causal claims.

5.4.2   Objection 2: ABM Lacks "Materiality"

At this point, an advocate of the potential outcome approach may object that there is still another fundamental difference between experiments and observational methods, on the one hand, and ABMs, on the other hand, which makes the former a superior tool for causal inference. The difference would depend on the "materiality" of the set-up studied with experimental and observational methods, as opposed to the abstract nature of the computer program investigated by an ABM. ABM's lack of "materiality" thus refers to the absence of real units of analysis concretely acting in natural or experimental settings. This materiality would allow experimental and observational methods to generate novel causal knowledge, something an ABM is structurally unable to do. As Diez Roux (2015: 101, emphasis added) put it:

> In the real world, we have *fact* (what we observe) and we try to infer the counterfactual condition (what we would have observed if the treatment had been different). In the simulated world, everything is counterfactual in the sense that *the world and all possible scenarios are artificially created* by the scientist.

The "materiality" argument initially appeared in the literature comparing numerical simulations in physics or economics and laboratory experiments (for detailed discussions, see Guala 2002; Winsberg 2003; Frigg and Reiss 2009: §5; Reiss 2011a). Given that an ABM is a special type of numerical simulation and the potential outcome approach aims at importing the experimental protocol within statistical methods for observational data, the materiality objection can easily be extended to the comparison between observational methods and ABMs, and deserves a dedicated discussion.

In particular, the materiality objection to using ABMs as inferential devices would unfold as follows. In observational methods, the novelty of the knowledge produced comes from letting the real system work as a data-generating mechanism, which, under experimental or quasi-experimental conditions, allows one to generate evidence that narrows down the class of admissible hypotheses. In contrast, an ABM is not made of the same "stuff" as the target it is designed to study. As a consequence, it simply cannot generate the kind of novel evidence needed to eliminate uncertainties on the nature of the mechanism, no matter how much theory/data are fed into them. The evidence provided by an ABM can at most expose the (deductive) implications of available knowledge. No adjustment in a fictional system can, by itself, provide evidence for any empirical hypothesis (for a similar point about models more generally, see Hausman 1992; Grüne-Yanoff 2009b). Thus, no matter the limitations of experimental and observational methods, the behavior of an ABM is simply not as relevant to its target as the behavior of a sample is to the behavior of the target population. The artificial nature of the system an ABM creates would make an ABM irrelevant by construction to speak about the real world.

The "materiality" argument is so appealing that, as surprisingly as it may seem, it can be found even among agent-based modelers. For instance, Gilbert and Troitzsch (2005: 13) remarked that, although "simulation is akin to experimental methodology", the two are not the same in that "while in an experiment one is controlling the actual object of interest (...) in a simulation one is experimenting with a model rather than the phenomenon of interest".

What now should be acknowledged is that a number of philosophical papers have recently challenged the materiality objection arguing that materiality is not enough to establish the superiority of experimental and observational methods over computer simulations (see, in particular, Barberousse et al. 2009; Parker 2009; Morrison 2015: ch. 6).

An important line of argument in this literature is that what really matters for allowing credible inferences is "whether the experimental and target systems were actually similar in the ways that are relevant, given the particular question to be answered about the target system" (Parker 2009: 493). The point is that the materiality of the system does not by itself guarantee this similarity. When experiments and statistical methods for observational data leave out important substantive elements like interaction structures, detailed time ordering of events, and dynamic and multi-level feedback loops, as they usually do compared to ABMs (see, for instance, Arnold et al. 2019), the fact these methods rely on a sample of real subjects, i.e. their materiality, does not make them *ipso facto* more reliable inferential devices than ABMs, which, although working only with virtual subjects coded by the computer program, typically do take into account all these aspects (for a similar remark applied to mathematical models *versus* ABMs, see Page 2008).

Thus, according to the critics of the materiality thesis, what ultimately must be assessed to judge the quality of the inferences made thanks to an ABM is "whether or not, to what extent, and under what conditions, a simulation reliably mimics the (...) system of interest" (Winsberg 2003: 115). The argument from materiality is, in this respect, question-begging (on this point, see also Morrison 2015: 243).

In the light of my discussion of ABMs with high "theoretical", "input", and "output" realism (see Chapter 3, Section 3.4), the reader can easily guess that I am going to follow this view here. I do believe that it is not material similarity in and by itself that grants the reliability of scientific inferences, but the (external) validity of the experiment, or the simulation, with respect to its intended target. Under particular conditions, an ABM can be taken as an approximate replica of its target (see Chapter 4, Section 4.1). In particular, when empirical data can be fed into at least part of an ABM's inputs through empirical calibration, and independent datasets can be used to assess the ABM's simulated outputs (through an operation that I called empirical validation), the lack of materiality of an ABM ceases to be a problem because the ABM's input and output realism makes it work as a "mimicking device", to borrow a metaphor from Mary Morgan (2012: 337), and, on this basis, it can be exploited to make inferences on how the mechanism modeled by the ABM operates in the real world .

As was the case with formal assumptions, materiality, too, is a red herring. Experiments and statistical methods for observational data have to face the same challenge as computer simulations. In both cases, formal and substantive assumptions have to be justified theoretically and grounded on empirical data that are external to the method. Materiality is neither sufficient (within experimental and observational methods) nor necessary (for ABMs) to ground the sort of similarity between the model and the target that is required for making the model a reliable inferential device (on the inescapable ultimate subjective nature of these judgments of similarity, see Sugden 2013).

5.4.3   Objection 3: ABMs Lack "Robustness"

Even if one agrees that substantive justification is required for the "formal" assumptions that experiments, statistical methods for observational data and ABMs all need to make credible causal inference in the presence of insufficient empirical data, and that the materiality of the setting under scrutiny does not assure *per se* greater credibility of experimental and observational methods, one may still object that, in the end, these methods grant credible inferences because, unlike ABMs, well-established procedures for testing the model's validity exist. In a recent comparison of DAG-informed regression modeling and ABMs, for instance, the observation was made that one of the strengths of the former is to "provide robust estimates of causal effects for clearly defined exposures and outcomes" whereas, within ABMs, "model complexity makes parameterization, calibration and validation difficult" (see Arnold et al. 2019: 251). Thus, ABMs would be weakened by a sort of structural lack of "robustness".

As previously noted, it is true that it is still an open issue in the ABM field to determine what counts as the best way of comparing simulated outcomes and empirical data describing the real-world outcomes the ABM wants to reproduce (see Chapter 4, Section 4.1.3). Still, with regard to the objection of an intrinsic lack of "robustness" of ABMs, I would like to argue that the procedures that experiments, observational methods, and ABMs exploit to convince a given audience that the inferences they produce are not artefactual are in fact qualitatively similar, and can be interpreted along the same lines.

As remarked by Parker (2008: 374), the aim of validity tests on simulations is, in analogy with statistical tests over experimental designs, to ensure that simulation results constitute persuasive evidence for some real-world hypothesis H. This, in turn, depends on whether "(i) the results fit H; and (ii) it is unlikely that the simulation study would deliver results that fit so well with H, if H is false". As a consequence, Parker (2008: 382) argued, simulation hypothesis testing is ultimately animated by an error-statistical approach according to which "(...) in order to claim that simulation results provide good evidence for some hypothesis of interest, we would be required to show that the potential sources of error were unlikely to have been present or to have impacted the results by more than a specified amount, rather than just that the evidence collected so far is consistent with their absence or their having minimal impact".

I explained in Chapter 4 that the toolbox agent-based modelers can draw upon to track and quantify these sources of error is made of a systematic combination of (i) an empirical calibration of an ABM's inputs; (ii) empirical validation of an ABM's outputs; and a variety of reliability checks performed through sensitivity, robustness, dispersion, and model analysis. Although I am well aware that there are practical difficulties in performing such tests for any ABM above a minimal level of complexity, my point is that it is unclear on what grounds one could deny that validity tests performed by experimentalists and users of statistical methods for observational data in fact follow the same principles, and claim that external validity is harder to establish in ABMs compared to those methods (for a claim of this sort with respect to simulations *versus* experiments, see, for instance, Morgan 2003: 231).

Obviously, I agree that, no matter how well calibrated and validated empirically, and tested through various reliability checks, an ABM is, it remains possible that the model lacks external validity. The mechanism postulated by the modeler may not capture the most relevant one. The data may also be the result of a plurality of mechanisms, whose operations and relevance for the phenomenon is simply masked by the postulated mechanism. No serious modeler can completely exclude that the knowledge provided by the simulation is insufficient to support causal inference. We have seen that the impact of certain non-calibrated formal assumptions concerning, for instance, functional forms can be assessed by sensitivity and robustness analysis. However, this may not be sufficient to persuade a given audience of the model's validity. Even when empirical data are rich, some of the ABM's substantive assumptions are likely to remain untested, thus limiting the validity of the inferences one can make on the mechanisms operating outside the world of the ABM.

But do not statistically oriented modelers, especially from within the potential outcomes framework, face analogous challenges? As discussed at length in Sections 5.2 and 5.3, some formal assumptions of experimental and observational methods can be only partially tested with empirical data—for instance, the relevance condition in IV estimators—whereas others cannot, by construction—for instance, the exogeneity assumption in IV estimators or, more generally, the meta-assumption that the causal graph under scrutiny is correct. Thus, background knowledge cannot completely exclude that the conclusion reached by means of experimental and observational methods is systematically biased or confounded.

That is why the most rigorous experimentalists and users of statistical models for observational data recommend performing systematic sensitivity analysis to assess to what extent empirical estimates from observational methods are robust against confounders (for an overview on the topic, see Gangl 2013: 385–90).

As an illustration of this, let me consider VanderWeele's (2011) re-examination of previous studies of the impact of being exposed, within a dyadic relationship, to someone who is obese or smokes on the probability of being also obese or smoking (the author also considered other outcomes but this is not relevant here). VanderWeele (2011), first, numerically represented various amounts of potential latent homophily within dyads that may have acted as a confounder for the interpersonal effect of interest; then, he determined what amount of latent homophily was necessary to make the originally observed interpersonal effect disappear; and, finally, he reasoned whether or not this level was realistic. This "robustness" check of the results of the original observational network studies is clearly based on transparent sensitivity techniques that can be considered "robust" but it is also clearly based on subjective judgments that must be defended on the basis of arguments that are neither technical nor empirical. And, indeed, VanderWeele (2011) honestly admitted in conclusion that "the approach we have described here is, however, subject to various limitations. Perhaps most importantly, in attempting to reason about whether latent homophily might explain away a contagion effect estimate we have had to make decisions regarding whether homophily of a particular magnitude is or is not plausible." And, Christakis and Fowler (2013: 568), whose results were at the heart of VanderWeele's sensitivity analysis, summarized the debates around their studies on putative contagions effects for various behaviors and outcomes with the following methodological remark: "In short, we believe that the key issue is the extent to which one can be explicit about one's assumptions, and the reasonableness of those assumptions, in work analyzing social networks as in any other statistical work." Interestingly, this reasonableness is emphasized with respect to sensitivity tools themselves. Gangl (2013: 399) observed indeed that these tools "would be degraded to little more than a computational exercise" in the absence of background empirical and theoretical knowledge suggesting the likelihood and extent of confounding. Substantive assumptions are necessary to the inferences, which in turn are not always verified or verifiable.

Let me now consider an example from the ABM field. In particular, Manzo and van de Rijt (2020) built an ABM of how a virus spreads through a population where the underlying structure of physical contacts between individuals has a broad degree distribution, i.e. one that includes a few high-contact individuals and a fat tail of individuals with relatively few daily social contacts. After calibrating the degree of the simulated network within the ABM through national representative contact survey data, Manzo and van de Rijt showed that the presence of hubs within the network is responsible for rapid viral propagation, and that immunizing the hub first mitigates the virus propagation more effectively than other strategies. However, one may argue, depending on how clustered dyadic physical encounters are in reality, a network feature for which Manzo and van de Rijt did not have direct empirical observations, the impact of preferentially targeting hubs may be over-estimated by the simulations (if network clustering is lower in the simulated networks than in real ones) or under-estimated (if network clustering is higher in the simulated networks than in real ones). Thus, to control for these unobserved network features, Manzo and van de Rijt (2020) re-ran the entire set of simulated interventions under various amounts of network clustering so that the interaction between the observed skewness of the contact distribution and the unobserved local clustering of these contacts could be fully mapped. Once this map was established, Manzo and van de Rijt could argue that, for a realistic level of clustering (according to the sparse existing evidence on clustering in the network of physical contacts), the originally simulated effectiveness of preferentially targeting hubs compared to other strategies was not canceled out.

Qualitatively, VanderWeele (2011) and Manzo and van de Rijt (2020) followed a similar procedure to quantify the uncertainty surrounding the putative causal effect of the network-related mechanisms of interest. The specific approach to sensitivity and robustness analysis may differ but the call for sensitivity analysis among users of statistical methods for observational data is qualitatively similar to the use of reliability tools adopted within the ABM field. Statistically oriented scholars, simulationists, and experimentalists all struggle with analogous issues of precision, accuracy, and calibration. In each methodology, a complex codified combination of background substantive knowledge, external empirical evidence, and sensitive and robustness procedures is necessary to eliminate uncertainty as regards possible sources of error.

5.5 A Common Logic: "Abduction"

To conclude, if one accepts that experimental and observational methods are fundamentally similar to ABMs with respect to the challenges they have to face in producing persuasive causal claims from a dependence and a production account of causality respectively, the question, then, is: On what grounds can the required combination of background substantive knowledge, external empirical evidence, and testing statistical procedures justify causal inference?

The justification, I suggest, may be rationalized by a pattern of reasoning known in the philosophy of science as "inference to the best explanation" (Lipton 2004). The central idea is that "the phenomenon that is explained (...) provides an essential part of the reason for believing the explanation is correct" (Lipton 2009: 629). In short, the hypothesis must be true, because if it were, it would provide the best explanation of the data. This line of reasoning seems to be at work among both quantitative scholars and ABM practitioners.

In the case of ABMs, the ultimate element that makes one believe that the match between the simulation of a well-calibrated generative hypothesis and the high-level patterns to be replicated is that this match is so "lovely"—to paraphrase Lipton—that it is rational to believe that the hypothesis is true (see Delli Gatti et al. 2018: 42). In turn, when this reasoning is at work, it makes the counterfactual scenarios where something changes in the simulated mechanism credible in reality, too. Analogously, in the case of observational methods, background knowledge in a given context makes it implausible that evidence of a robust dependence is not due to a causal relation, which is at the same time responsible for the different responses under different hypothetical circumstances.

In both cases, the reasoning goes as follows: the explanation that the truth of the causal claim provides for the data is so "lovely" that it is rational to believe that the causal claim is true. Its circularity notwithstanding, this reasoning is, I suspect, ultimately operating in the case of simulated outcomes (i.e. ABMs) as it is in the case of potential outcomes (i.e. experimental and observational methods). Ultimately this may be the case because, as suggested by Sugden (2013), this form of abduction is probably operating at the psychological level in any form of modeling exercise that has been proven to be able to produce some regularity that replicates *to some extent* corresponding regularity in the real world.

6

Method Diversity and Causal Inference

In the last two chapters, I studied the conditions under which agent-based models (ABMs) (Chapter 4) and methods relying on experimental or observational data (Chapter 5) can produce persuasive causal claims. The main argument that I tried to defend was that both types of methods ultimately rely on assumptions that, in some cases, are untestable *in practice*, and, in other cases, are untestable *in principle* so that empirical data are insufficient *alone* to ground the credibility of causal inferences within these methods. Consequently, I argued, ABMs as well as experimental and observational methods can only support causal claims through a complex mix of empirical data, theoretical arguments, and reliability checks.

This analysis was premised on the conclusion of Chapters 1 and 2 where I explained that ABMs, experiments, and statistical models for observational data think about mechanisms and causality in different ways, and, to grasp these differences, I proposed the distinction between a vertical view on mechanisms squaring with a production account of causality (in the case of ABMs) and a horizontal understanding of mechanisms squaring with a dependence view on causality (for experiments and observational methods). The argument I progressively built through the two previous chapters was based on the assumption that these different ways of thinking about mechanisms and causality were self-sufficient. I accepted this simplification to provide the most "honest" assessment of the conditions under which each method can lead to persuasive causal claims, i.e. an assessment that respects the specific accounts of mechanisms and causality that mostly cohere with each method's inner logic and functioning.

In this final chapter, in order to draw all the implications from my argument that ABMs, experiments and observational methods face qualitatively similar challenges to establish causal claims along *their own view* on mechanisms and causality, I go a step further and, rather than accepting these specific perspectives at face value, I argue that both approaches to causality and mechanisms are in fact insufficient to sustain persuasive causal claims when operating separately.

In the light of the discussion of Chapters 4 and 5, I would like to suggest indeed that behind the necessity of both ABMs and experimental/observational methods to rely on assumptions lies a common problem, i.e. the inescapable pervasiveness of "systematic biases". I use the term here to refer generically to "any structural association between treatment and outcome that does not arise from the causal effect of treatment on outcome in the population of interest" (Hérnan and Robins 2020: 78). Within experimental and observational methods, systematic biases most often consist in confounding, i.e. the presence of a potential common cause for the treatment and the outcome (Hérnan and Robins 2020: ch. 7), "effect modification", i.e. possible changes of the values and/or the sign of the effect of the treatment within sub-groups of the population of interest (Hérnan and Robins 2020: ch. 4), or "mediation", i.e. the presence of intervening variables absorbing part of the putative causal effect through indirect paths (see Makovi and Winship 2021). Within ABMs, "systematic biases" take the form of mechanisms that, while at work in the real world, are not represented within the computer simulation, thus leading to over-emphasize effects that are smaller outside the world of the model or, vice versa, mechanisms that, while at work within the ABM, are of minor importance in the real world, thus leading to mechanism masking. In both families of methods, the deepest sources of "systematic biases" are similar, namely limited knowledge of the social world leading to the omission of certain variables/mechanisms and lack of data concerning known confounding, modifying, and/or intervening variables (Hérnan and Robins 2020: 92, 168–9). Measurement errors (Hérnan and Robins 2020: ch. 9), methodological mistakes—like conditioning on common effects of independent causes, thus opening collider paths (Elwert and Winship 2014)—or random variability (Hérnan and Robins 2020: ch. 10) can also generate systematic biases. In both ABMs and experimental/observational methods, the consequences of "systematic biases" is the same: the causal interpretation of a given effect, be it vertical (within ABMs) or horizontal (within experimental and observational methods), will be attributed to a phenomenon which is not the "right" one.

That is the main reason, I will argue in this chapter, why it would be profitable to shift the debate from defending the superiority of particular methods, and thus of specific understandings of causality and mechanism, to discussing how ABMs, experiments and observational methods can provide a variety of evidential sources that compensate for each other's weaknesses. These methods should be integrated: they provide different type of data and arguments that can be used to reduce the uncertainty surrounding any causal claims because of the inescapable pervasiveness of "systematic biases". This is the core of the "evidential variety" thesis I want to push in conclusion of my analysis.

To defend the argument, I will go through five dialectical steps. First, I argue that acknowledging the legitimacy of a variety of methods for causal inference should not lead to radical forms of causal pluralism but rather prompt the question of how scholars with different methodological orientations can still fruitfully communicate about causation (Section 6.1). Second, to answer this question, I endorse a pragmatist theory of evidence that, I maintain, helps us to understand how causal claims are defended *in practice* by the scientific community (Section 6.2). Third, I elaborate on this view, and argue for "evidential pluralism", the thesis according to which methods that produce data and arguments of a productive kind must be integrated with methods that lead to data and arguments of a dependence kind, and rationalize the role of ABMs in this methodological synergy (Section 6.3). Fourth, I qualify this claim by explaining under what conditions a variety of evidential sources are effective for causal inference and why ABMs and experimental/observational methods satisfy these conditions (Section 6.4). Finally, I illustrate my plea for "evidential pluralism" by discussing specific examples of method synergies where experiments, statistical methods for observational data, and ABMs are explicitly confronted on the same causal issue with the aim of generating diverse but complementary knowledge for the causal claim of interest (Section 6.5).[1]

6.1 Causal Pluralism, Causal Exclusivism, and Evidential Pluralism

According to a common distinction among philosophers of science (see Williamson 2006), views on causality can be "monistic"—if one essentially believes that there is one notion of cause and one fundamental type of indicators of causality—or "pluralistic"—if one essentially believes that causality can be understand in different ways that are difficult to reconcile. The latter thesis is often referred to as "conceptual pluralism" (Reiss 2011b: 908) and, as we have seen in Chapter 1, is endorsed, among others, by Hall (2004: §6) who introduced the distinction between *dependence* and *production* accounts of causality on which I built in the previous chapters to compare experiments and statistical models for observational data to ABMs (on Hall's causal pluralism, see also Longworth 2006: §3).

Until now, my argument was clearly framed from within a perspective of conceptual pluralism on causation. I emphasized indeed that quantitative social scientists should recognize more clearly that, when they program ABMs, design experiments, or estimate statistical models on observational data, they in fact rely on different understandings of causation (and mechanisms). Acknowledging this variety of views, I argued, was the first step properly to assess under which conditions ABMs and experimental/observation methods can be used for causal inference *in their own terms*, meaning coherently with the view on causation that animates them (i.e. a production and dependence view, respectively) (see Chapter 1). One could then read as a justification of the view that there are distinct and equally legitimate notions of cause my subsequent analysis showing that causal claims from within a production and dependence view can in fact be supported by ABMs and experimental/observational methods, respectively, even though, in both cases, this result can only be achieved through a complex mix of data, arguments, and reliability checks (see Chapters 4 and 5).

Now, while I do believe that it is crucial to keep in mind that causation can be understood in various ways, and that different methods provide support for different indicators of causality, what I want to stress at this point of the analysis is that one should avoid drawing the wrong implications from conceptual pluralism on causality.

In particular, causal pluralism and the specific argument on ABMs and experimental/observational methods I defended from within this perspective, could be used to motivate users of a given notion of cause (and a given method) to claim an exclusive authority on how their own notion of cause ought to be used. By the same token, though, they would lose any authority on how the notion of cause is used in other communities. Worse still, seemingly appropriate disagreements on causal knowledge would turn out to be, on reflection, inappropriate. Since the meanings of the concepts involved are different, causal

---

[1] Sections 6.1–6.4 of this chapter build on and extend Casini and Manzo (2016: 56–62).

claims on the same matter, but formulated by different communities, which operate with different standards through different methods, could not contradict one another. They would rather be orthogonal to one another.

This attitude, which one may call "causal exclusivism", is a drawback of causal conceptual pluralism that should be avoided. Causal exclusivism is indeed normatively detrimental, insofar as it would, if endorsed, obstruct critical analyses of the limitations of each method for causal inference and thus impede the search for synergies that aim to compensate for them. Causal exclusivism seems also descriptively inaccurate, insofar as it clashes with the observation of how scientists from different communities *de facto* react to each other's results. It is certainly true that sometimes scientists talk past each other or ignore each other's results merely on the ground of the method that was used to obtain them. However, scientists also treat each other's results seriously, even when obtained by methods they do not favor, and by calling for further tests by other methods, they accept that previously endorsed conclusions may be retracted. A recent example of this typical dynamic of science is given by studies that aim at reproducing results on SARS-CoV-2 propagation, and possible intervention policies, generated by traditional compartmental models in epidemiology by using ABMs based on different formal and substantive assumptions (see, for instance, Chattoe-Brown et al. 2021).

In fact scientists' attitude towards data, methods, and the results of others' studies seems well described by a form of "default entitlement", according to which scientists trust the evidence produced by other scientists unless there are domain- or case-specific reasons to distrust it or challenge it (Reiss 2015: 354–5). If one accepts this perspective, then causal pluralism should not lead to causal exclusivism but to the less radical view that different understanding of causation, rather than being associated to separated methodological islands, comes along with various types of indicators of causality, and corresponding types of data and argument potentially supporting these indicators, a perspective that can be labeled "evidential pluralism". Russo and Williamson (2007: 168) nicely characterized this perspective when, to summarize how causal claims are typically defended in bio-medical studies, they observed:

> In the health sciences it is clear that there are a variety of kinds of evidence. For example, in cancer science one might have a dataset containing clinical observations relating to past patients, another containing observations at the molecular level, some knowledge of the underlying biological mechanisms, some knowledge about the semantic relationships between variables provided by medical ontologies, and so on. All these types of evidence will shape causal beliefs about cancer: the datasets provide statistical evidence concerning difference-making; mechanisms provide evidence of stability; semantic relationships may provide evidence against causal connection (two dependent variables that are ontologically but not mechanistically related do not require a causal connection to account for their dependence).

Compared to both causal pluralism and exclusivism, "evidential pluralism" seems a fruitful position because it only requires that there be a communication "space" where, irrespective of one's preference towards a particular method for causal inference, mutual understanding and genuine disagreement are possible. According to "evidential pluralism", this communication "space" concerns the way a certain type of data and arguments has been produced that can speak in favor or against a given causal claim, and scientists deliberate on how much support these data and arguments provide for the claim (see Casini 2012).

Fully to defend the "evidential pluralism" perspective, three questions must be answered. First, what kind of data and arguments is worth collecting? Second, what gives to data and arguments a warrant-increasing value, despite the apparent difference between the causal views that different researchers subscribe to? Finally, why do different types of data and arguments actually matter? Reiss's (2015) pragmatist theory of evidence can be used to answer the first two questions (Section 6.2); the literature on evidential pluralism itself helps to answer the third one (Section 6.3).

6.2 A Pragmatist Account of Evidence

Reiss's (2015: 341) pragmatist theory of evidence is relevant for my argument because Reiss's contribution is motivated by rejection of the view that there would be a gold standard for causal inference, namely "randomized experiments". According to Reiss (2015: 349), this belief would arise from a common fallacy of monistic views of causality that "mistake evidence for whether or not a causal relation is present for the relation itself or for constituting the meaning of causal claims". Thus Reiss's starting point is very much in line with the present book's general ambition, which is to question the widespread beliefs among quantitative social scientists that experimental and observational methods, and the dependence view on causation embedded in these methods, in fact represent the most powerful way to produce data and arguments supporting causal claims. Similarly to Reiss, my intention, too, is to ask experimentalists and users of statistical methods for observational data to consider that one thing is one view on what causality is and another thing is what constitutes data and arguments for building persuasive causal claims.

As to this specific question, i.e. the kind of knowledge worth collecting to defend a causal claim, Reiss's (2015: 349) pragmatist theory of evidence posits that there are several (well-known) kinds of facts one is (defensibly) entitled to expect under the supposition that a causal claim is true. These facts are:

(1) the obtaining of correlations between putative cause and effect;
(2) the obtaining of changes in the effect after interventions on the cause;
(3) the fact that the cause is a necessary or sufficient condition for the effect;
(4) the existence of a mechanism for the relation;
(5) the existence of a continuous process triggered by the mechanism that leads from the cause to the effect.

According to Reiss, irrespective of one's convictions on what causality essentially is, studies that produce data and arguments for these facts provide evidence for causality in two possible ways. First, they may directly support a causal hypothesis. Second, their results may be incompatible with what one is entitled to expect under the supposition that some alternative hypothesis is true, thereby indirectly supporting the causal hypothesis (Reiss 2015: 352). That the evidence may be collected by using certain methods is, for Reiss (2015: 358–60), a contingent fact, of no deep significance for the justification of causal inference. What matters is how the data and arguments produced by this or that method contribute to the plausibility of the causal claim in relation to relevant alternatives.

In the context of my analysis, one may paraphrase the above proposal as follows. What matters to causal inference is generating knowledge that bears on assumptions, whose truth would allow one to safely draw the inference from data. In particular, the warrant for the inference comes from data and arguments that can increase the credibility of assumptions under which the causal claim is plausible or decrease the credibility of assumptions under which it is implausible. This, by itself, does not entail that one should collect data and build arguments by using one method rather than another. In fact, the assumptions that would warrant the reliability of a method are among those very facts one would need to establish prior to deciding to use the method to generate data and arguments. Motivating those assumptions requires further data and arguments, which requires further assumptions, and so on.

This leads to the second question raised by "evidential pluralism": what in fact confers to a specific piece of information the capacity to increase the persuasiveness of a causal claim?

By reference to the case of the causal effect of smoking on cancer, Reiss (2015: 355–7) suggested that a causal claim becomes credible when a number of facts, expectable under the supposition that the claim is true, are together sufficient to convincingly rule out alternative explanations of those facts. For instance, it is known that smoking is strongly correlated to cancer, that more frequent smokers have a higher risk of developing cancer, that stopping has a beneficial effect, that cancer develops sometimes after taking up smoking, etc. Thus, it is plausible to conclude that smoking (rather than, say, some genetic factor) causes cancer.

Clearly, this is a case of "inference to the best explanation", or abduction, in the sense introduced at the end of Chapter 5 (see Section 5.5) where I argued that, in fact, both ABMs and experimental/observational methods ultimately build on this type of circular reasoning to defend the reliability of the knowledge they produce. In particular, the truth of the causal claim is inferred, because it provides the best explanation of the data among a pool of alternative hypotheses, by either being directly supported by data and arguments, or by being indirectly supported by data and arguments that rule out the alternatives.

Naturally, disagreement can arise on how one should move from collecting data and developing arguments to defend a causal claim, especially when the results of studies that follow different methods pull in different directions. Different communities may prefer data and arguments coming from different sources. As I have shown in the two previous chapters, ABMs produce knowledge of one kind, based on generative mechanisms, whereas experiments and observational methods produce knowledge of other kinds, based on other criteria, which are in one way or another based on a difference-making intuition. However, I share Reiss's conviction that favoring a specific method for causal inference does not amount to a justification for believing that only one kind of data and arguments is relevant for establishing a causal claim. In connection with randomized control trials (RCTs), Reiss (2015: 358) noted:

> Experimentalists are methodological foundationalists. They believe that some results are produced by methods that are intrinsically reliable and therefore epistemically basic. The epistemically basic method is a well-designed and well-executed randomized experiment. (...) The account proposed here instead begins with

the hypothesis and inquires about what kinds of facts we need to collect or learn in order to be entitled to infer the hypothesis. That these facts can sometimes be learned in an experiment is trivially true but, according to this account, contingent and of no deeper significance for the justification of inferences.

This observation should be extended to all methods I studied in this book, including ABMs. The perspective that I would like to push is one where what matters to causal inference is not so much the methodological source of the data and the arguments but how these data and arguments increase one's confidence in some relation not being affected by "systematic biases", given constraints due to limited data and uncertain background assumptions. Various kinds of knowledge may be relevant in different ways. This is the basic intuition at the heart of "evidential pluralism", which I develop next.

6.3 Evidential Pluralism and "Coherentism"

Russo and Williamson's (2007) "evidential pluralism" thesis was motivated by the empirical observation that, in the context of biomedical studies, causal inference is recurrently established *in practice* through various combinations of data and arguments of both a difference-making and production kind (for a recent extension and confirmation of this observation, see Williamson 2019). Observations supporting the intuition that "evidential pluralism" may operate in other fields have been made in macroeconomics (Claveau 2011, 2012) and econometrics (Moneta and Russo 2014). Specific research topics in epidemiology have also been investigated from the point of view of "evidential pluralism" leading to a new interdisciplinary extension of the thesis according to which "some disciplines may provide evidence for the probabilistic angle of causal inference, whilst others may provide the productive one" (Canali 2019: 4).

But why different type of data and arguments could be necessary to defend persuasively a given causal claim? The justification for evidential pluralism is that the uncertainty about the causal nature of a given relation between two happenings may be reduced in (at least) two ways. On the one hand, data and arguments on credible entities, activities, and interactions—i.e. in Chapter 1's terminology, a mechanism in a vertical sense—influences one's belief that some relation is genuinely causal. Intuitively, empirical evidence of robust relations not backed by data and arguments on the entities, activities, and interactions responsible for them does not eliminate the doubt that the relation of interest could disappear, or be weakened, because of unobserved confounders, effect modifiers, and/or intervening variables. As Illari (2011) neatly put it:

> If you have evidence of a mechanism linking C to E—evidence of entities, activities and their organization in the right place at the right time—you have some reason to believe that there is a causal relation between C and E. This is because you have traced the mechanism linking C to E. (...) Take protein synthesis again. Once you have found the entities and understood what they do—their activities—and how they are organized relative to each other and the cell, you can trace the path from DNA through replication, to transcription and the various forms of RNA and their activities, to the ribosome and the creation of the amino acid chain. Since you have

traced the causal process that begins with the DNA and ends with the required amino acid chain, you can be confident that one can begin with DNA and end up with an amino acid chain.

However, on the other hand, empirical evidence of a robust relation—i.e. in Chapter 1's terminology, a mechanism in a horizontal sense—also influences one's belief that the relation may be causal. Intuitively, data and arguments on how given entities, activities, and interactions combine to produce some behavior actually does not guarantee that their organized operation is not masked in the broader context of further (possibly unknown) modes of interaction and outer influences so that the overall outcome of the postulated mechanisms *is not* the expected putative causal relation of interest. As Illari (2011) again precisely put it:

> The general problem for inferring 'C causes E' from evidence of a mechanism link-ing C and E is the problem of masking. You have found one link from C to E, but you do not know what other links there may be. (...) This means there may be many cases where we can trace a mechanism between variables, but still have no idea whether the first variable increases, decreases, leaves untouched, maintains in homeostasis, and so on, ... the other. (...) Complementary difference-making evi-dence is required.

Thus, if one accepts this justification of "evidential pluralism", it becomes clear that data and arguments on mechanisms understood as a multi-level dynamic system of interacting entities and activities are not more fundamental than data and arguments along depend-ence and making-difference lines, or vice versa: rather, the two kinds of knowledge com-plement each other. Once more, as Illari (2011) effectively put it:

> evidence of a difference-making relation between C and E, and evidence of a mech-anism between C and E, are not just independent 'pluses' in favour of the causal claim 'C causes E'. This is because evidence of difference-making and evidence of mechanism integrates: each addresses the major weakness of the other evidence. Together they are much better evidence for the existence of a causal relation than evidence either of difference-making, or of mechanism, can be on its own.

But how could this thesis be framed in relation to data and arguments produced by ABMs, on the one hand, and experimental and observational methods, on the other hand? The strength of the knowledge produced by an ABM resides in its power to reduce the uncer-tainty on whether the source of some observed relation at a given level of analysis, usually described through experimental and observational methods, can actually be reproduced through the action of a well-specified mechanism. In principle, a careful use of the method, along the lines sketched in Chapter 4 (Section 4.1), can offer novel knowledge for causality on connections across different levels (or scales) of analysis, given the simulated mechanism. In practice, as pointed out in Chapter 4 (Section 4.2), such a use is hampered by data and knowledge limitations, which may not be fully counterbalanced by the *from-within-the-method* reliability tools described in Chapter 4 (Section 4.3). In such cases, experimental and observational methods can surely help. Indeed, ABMs make use of empirical knowledge got from these methods to formulate assumptions, calibrate the model's inputs, validate the model's simulated outputs, and, more generally, to modify or refine a previous model.

At the same time, although empirical knowledge coming from experimental and observational methods is useful for ABMs, this does not make an ABM's generative knowledge dispensable. When the assumptions of experimental and observational methods that are necessary to handle potential confounding and other systematic biases are not well confirmed by data or justified by domain-specific considerations—which is likely to be constantly the case, as I have shown in Chapter 5—, ABMs allow one to model those scenarios over which data are missing, thus providing a further protection against possible violations of those assumptions. In this sense, from the point of view of experimental and observational methods, an ABM can be thought of as a theory-driven sensitivity technique. It can be used to design a variety of mechanisms, making different hypotheses on different sets of potential confounders, effect modifiers, and/or intervening variables and assess under which of these mechanisms the expected (or observed) correlations persist (for some examples of this, see Section 6.5). In this way, an ABM can increase one's confidence that a relation of interest is not affected by systematic biases. A theoretically informed, empirically calibrated, and validated ABM would thus provide a severe test for the existence of a given observed dependence relationship (for the view of "statistical inference" as severe testing, see Mayo 2018).

In sum, ABMs and experimental/observational methods provide necessary and complementary knowledge for building persuasive causal claims. On the one hand, an ABM adds specificity to experiments and statistical models for observational data in that it formalizes, and deduces consequences from, specific sequences of events that may have generated the dependence relationship identified by those methods, in particular when the latter's assumptions are hard to realize or difficult to check. On the other hand, experimental and observational methods add plausibility to ABMs in that they provide empirical information that constrains the modeler's theoretical creativity on both input and output side. Thus, the thesis of evidential pluralism ultimately fosters a virtuous circularity between the two methods.

To conclude, let me finally note that, seen this way, "evidential pluralism" ultimately leads to framing causal inference through the lenses of a "coherentist" approach to epistemic justification. Within this perspective, indeed, the defense of a given causal claim is no longer sought in one specific type of information that would be more fundamental than another (data *versus* theory, for instance) or in one specific property that would constitute the ultimate sign of causality (manipulability *versus* production, for instance). In contrast, the credibility of a given causal claim would arise from the degree of coherence among a set of beliefs on various pieces of information. Various "coherence" theories of epistemic justification exist, and one of the main results of the lively debate in this field is clearly that no consensus seems to exist on what "coherence" actually means (for a review, see Olsson 2017). But, for the purpose of my argument, what matters is the general point of view of "coherentism", i.e. the intuition that the "web of beliefs" that one can construct to defend a given proposition makes the justification for holding the proposition. In Olsson's (2017: 3) words, "(...) the fact that our beliefs cohere can establish their truth, even though each individual belief may lack justification entirely if considered in splendid isolation, or so it is thought". Applied to the problem of causal inference, this intuition implies that no specific piece of knowledge should be seen as playing a special role in persuading a given audience that the causal claim is defensible. The persuasiveness of the claim would rather be contingent on the way one is able to combine insufficient data, theoretical justifications of untestable assumptions, and results from reliability tools (like sensitivity and robustness analyses), plus the combination of all these pieces of knowledge from various methods. It is the degree to which all these pieces of information cohere that will make the claim more or less convincing in the eyes of this or that audience. As noted by Sugden (2013) in connection with abstract models in economics, there will always be a subjective component in these judgments. However, multiplying the pieces of information that one has to mobilize to defend the coherence of the whole system of clues, as I am suggesting here, should increase the number of constraints to be satisfied, thus reducing the subjective character of the justification exercise (for a formal treatment of the conditions under which this intuition may be correct, see Landes 2020).[2]

6.4   When is Diverse Evidence Most Relevant?

My plea for "evidential pluralism" applied to the relationship between ABMs and experimental/observational methods requires a further qualification, though. After all, the two classes of methods may produce data and arguments on different aspects of a given dependence relationship but these data and arguments may be affected by qualitatively similar systematic biases, errors, and random variability. Under what conditions then is the plurality of data and arguments produced by ABMs and experimental/observational methods relevant to causal inference?

It seems reasonable to postulate that diverse data and arguments are the most effective when one can regard as "independent" the particular set of formal and substantive assumptions on which are based the methods exploited to produce these data and these arguments. "Independent" essentially means here that the assumptions on which a given model is based focus on different aspects of the reality compared to another model, thus introducing different simplifications. For this reason, models based on "independent" assumptions are likely to introduce different biases and make different errors. Thus the basic intuition is that independent sources provide more "diverse" data and arguments because they are affected by different types of error; and if these diverse data and arguments agree, then they produce knowledge that is more confirmatory (Bovens and Hartmann 2003: 96–7). This is the main idea behind the many-models approach theorized by Scott Page (2018: 28) who suggested seeing diverse models as "independent lies".

Although differently phrased, the argument seems present in many modeling exercises. For instance, in Chapter 3, I mentioned Ajelli et al.'s (2010) comparison of a data-driven ABM of virus propagation with a more parsimonious compartmental model, a comparison that the authors concluded by observing that "The good agreement of the two

---

[2] I should thank Daniel Little for pointing me to "coherentism" in the philosophy of science and suggesting a "coherentist" interpretation of the "evidential pluralism" thesis.

approaches reinforces the message that computational approaches are stable with respect to different data integration strategies and modeling assumption". With reference to epidemiological studies of obesity, Marshall and Galea (2015: 97), although regretting the lack of model comparisons on this topic, made a similar observation concerning the potential value of the diversity of assumptions underlying different types of models:

> (...) currently missing from the literature are comparative studies in which investigators interrogate an epidemiologic question with different types of causal inference models, including those that are agent-based. These may fruitfully be the focus of future work. [...] Constructing and calibrating an ABM with these same data would permit direct comparisons of the assumptions made by each method and would also reveal specific situations in which the ABM approach may provide novel and important public health insights to curb obesity.

More recently, Chattoe-Brown et al. (2021: 9, 16, 18) have attempted to reproduce, through an ABM, the published results of an influential compartmental model of the SARS-CoV-2 epidemic in the UK, and observed:

> Our agent-based model implementation (the Reproduction Model, RM) differs from the LSHTM Target Model (TM) in several ways. Many of these reflect the different styles of modelling involved (compartmental versus ABM) and therefore if we can still reproduce the policy outcomes of the TM using the RM, this would support the contention that these might be substantive behavioural outcomes and not mere artefacts of the modelling approach. (...) The reproduction process provides benefits for both the TM and the RM. If the RM reproduces the TM, then it adds independent credibility to the TM's policy prescriptions. If the RM does not reproduce the TM, it warns us that the TM may be erroneous (at least based on the information about it that has been made readily accessible). (...) Reproduction, which can be attempted using different modelling approaches (as here with an ABM reproducing a compartmental model) can give insight into the robustness of model results based on assumptions that are not questioned or possibly cannot be relaxed within a particular modelling approach.

More precisely, as argued (in a Bayesian framework) by Schupbach (2015), the role of diverse data and arguments is to make it less plausible that a robust result is explained by any alternative hypothesis than it is by the hypothesis at stake. Accordingly, the kind of diversity that is required for data and arguments supporting a robust result to be confirmatory is "eliminative diversity": data and arguments are (more) eliminatively diverse if they are (more) capable of ruling out salient competitors of the hypothesis. Methods based on (partially) diverse assumptions are more likely to produce this sort of data and argument diversity. As Schupbach (2015: 314) observed (with reference to the application of the Volterra principle to population dynamics):

> By utilizing these (...) diverse models, we rule out confounding hypotheses pertaining to our result left standing by any subset of the models used alone. Notably, we alleviate worries that our result is an artifact of a simplifying assumption common

to some subset of our models by duplicating that result using a new model that does not share that assumption.

It is precisely this qualified sort of "independence"—i.e. (partial) independence among background assumptions—that makes it possible that the combination of ABMs, experiments, and statistical methods for observational data increases the persuasiveness of a causal claim. Although in each of these methods assumptions are made that in practice are not fully testable, if the facts that one is entitled to expect under the supposition that the causal effect is "real" are reproduced by a model that does not make those untestable assumptions (albeit perhaps making other, untested and untestable assumptions), one is more justified in holding the causal claim to be provisionally "true".

In general, there is no reason to believe that the assumptions of an ABM—about how agents behave and interact and how such behaviors and interactions generate aggregates of interest—are going to be systematically related to the assumptions of experimental and observational methods—about how to causally interpret observed patterns among such aggregates—in such a way that the overall set of data and arguments produced is not eliminatively diverse. On the contrary, as shown, for instance, by specific comparisons between directed acyclic graph (DAG)-informed regression modeling and ABMs, the latter tends to focus on those aspects (like feedback loops and interferences or spillover effects) that are precisely excluded by construction by the former (Arnold et al. 2019: 2, 6, 9). Thus, we have reason to believe that the two families of methods are likely to work under sets of assumptions that are sufficiently independent to produce data and arguments that are eliminatively diverse in the required way. This is what ultimately justifies my plea for making ABMs and experimental/observational methods friends, rather than foes, when causal inference is at stake.

6.5 Examples of Method Synergies

To put this argument in context, I would like to conclude my plea for "evidential pluralism" by discussing some pieces of research where experiments, statistical methods for observational data, and ABMs are explicitly confronted on the same causal issue, and the possibility that these methods produce data and arguments along horizontal and vertical lines on the basis of different assumptions, thus generating diverse but complementary knowledge for the causal claim of interest, is explicitly discussed.[3]

6.5.1 Obesity: ABMs and Regression Models

My first example comes from scholarship on individuals' health-related behaviors and states that may be subjected to interpersonal influences and that have been studied through survey-based, thus observational, data that were not natively collected with proper data collection methods for relational data. On this topic, studies by Christakis and

---

[3] I should thank Christopher Winship for pushing me to look for concrete illustrations of my argument about how causal inference can benefit from method triangulation.

colleagues relying in particular on a longitudinal epidemiological cohort known as the Framingham Heart Study attracted a singular amount of attention (for an overview, see Christakis and Fowler 2013). In short, after careful data cleaning and coding of information on various types of social ties contained in this cohort, Christakis and associates discovered two main results: (i) that many health-related outcomes, including smoking, alcohol consumption, obesity, or emotions and mental states like happiness or depression, tended to cluster among individuals that shared some type of social ties with a higher frequency than expected by chance alone, and (ii) that, for most of the outcomes, this association between *ego*'s states and the states of her contacts tended to become statistically insignificant beyond three degrees of separation (for a summary, see Christakis and Fowler 2013: fig. 1). To establish the putative causal effect of *ego*'s being in contact with, say, a friend who is obese on *ego*'s probability of also being obese, Christakis and colleagues relied on regression models for longitudinal data specified at the dyadic level, including *alter*'s lagged independent variables (Christakis and Fowler 2013: §5). This specific feature was introduced explicitly as a possible control for homophily between *ego* and *alter* because, the authors argued, conditioning on *alter*'s lagged observed trait of interest may help to create independence between *ego*'s current state on the trait of interest and her state when the tie between *ego* and *alter* was created (Christakis and Fowler 2013: 566).

Christakis and associates were indeed perfectly aware that this was the main problem they had to overcome, i.e. to control for the possibility that *ego* and *alter* were already similar before they formed a tie and actually formed a tie precisely because they were similar. In that case, selection rather than social influence would explain the observed clustering of health-related behaviors and outcomes. For my purpose, the interesting point is the way Christakis and colleagues responded to this problem and to the various critiques their regression modeling approach received (Christakis and Fowler 2013: 561, 567). In particular, on the one hand, they emphasized the importance of making assumptions explicit, and, on the other hand, they defended the idea that confidence in causal inference can be increased by addressing the same causal issues through methods based on different assumptions. In their own words (see Christakis and Fowler 2013: 573, italics added):

> A key consideration, therefore, is what the standard for evaluating our findings is. Is the real issue whether such interpersonal influence for these interesting phenomena (obesity, emotions, etc.) occurs? In that case, confirmatory *work of various types by various investigators* should be taken to support our findings. (...) Or is the key issue here that interpersonal effects are hard to discern with confidence, and that data and methods are imperfect and subject to *assumptions* or *biases*? If so, we quite agree. This is one of the reasons we have tried to be *transparent about the methods* used in our work. This is also one of the reasons that we ourselves, and others working collaboratively with us, have proposed new approaches, such as *experiments* (both offline and online) and *instrumental variable methods* involving genes as instruments, both of which might be able to provide *different sorts of confidence* in causal inference. Here, the standard is whether an accurate observation is scientifically possible. We think it is.

In Chapter 5 (Section 5.5), I have already discussed one specific approach that was adopted to increase the confidence in Christakis and colleagues' estimates, namely the sensitivity analysis performed by VanderWeele (2011) based on making variable the possible amount of latent homophily that may have (at least partly) absorbed the putative causal interpersonal effect of interest. In the terminology of the previous section, however, VanderWeele's robustness check has a relatively low level of "eliminative diversity" because the method used to perform the sensitivity analysis was based on assumptions that were similar to those of the method whose results must be tested (VanderWeele 2011: 253). Indeed the approach to sensitivity analysis adopted essentially amounted to postu-lating the presence of a given latent variable (say, pleasure to enjoy food), numerically represent the prevalence of this factor within dyads where both *ego* and *alter* are, say, obese and within dyads where *ego* is obese but *alter* is not, and finally postulate a certain impact of the presence of the latent factor (given its postulated prevalence) on the proba-bility of *ego* being obese (VanderWeele 2011: 244–45, 248).

What I want to stress now is an even more severe test to which Christakis and col-leagues' estimates were submitted, i.e. an ABM developed by Zachrison et al. (2016) with the explicit intent to assess the capacity of the regression modeling approach pro-posed by Christakis and colleagues to identify the causal effect of interpersonal social influence when attribute homophily is present. Interestingly, Christakis and Fowler (2013: 570) also regarded this ABM as a complementary approach to assess the robust-ness of their results.

Differently from VanderWeele's sensitivity analysis where the role of a possibly unob-served variable generating homophily was only represented indirectly through its poten-tial effect, Zachrison et al. (2016: 2) designed an explicit process to create longitudinal network data of exactly the same kind as those studied by Christakis and colleagues but where the role of network influence and homophily were perfectly known because the mechanisms were explicitly coded within the ABM. In particular, Zachrison et al. (2016: 3–4) considered the case of obesity and represented a population of artificial agents where each of them had (i) a specific weight, (ii) a specific rate of weight gain, and (iii) the pos-sibility of creating ties with a tunable number of other agents (i.e. virtual friends). In the ABM, an agent is selected to create a friendship tie either at random (no homophily condi-tion) or proportionally to the reciprocal of the weight difference between *ego* and the potential friend (homophily condition). As to social influence, an agent's weight gain can depend on either her specific rate of weight gain *only* (no network influence condition) or a "tunable weighted average" between the agent's specific weight gain rate *and* the differ-ence between the agent's current weight and her friend(s) weight (network influence con-dition). Given these homophily and network influence mechanisms, Zachrison and colleagues estimated the regression models specified by Christakis and associates on the network data generated by the ABM and found that: (i) Christakis and associates' longi-tudinal regression model was able to detect correctly the presence or the absence of the network effect, regardless of the presence or the absence of the homophily mechanism within the ABM (see Zachrison et al. 2016: tables 1, 2, and 3); (ii) in contrast, the statistical model was not able to detect correctly the presence/absence of homophily, no matter whether the network influence mechanism was present or absent from the ABM (see Zachrison et al. 2016: tables 4 and 5). In other words, Christakis and colleagues' statistical model was highly sensitive and specific in identifying the putative causal interpersonal effect on a given outcome but it could not also reliably tell us if this happens in the presence or in the absence of homophily.

Thus, generating network data from the bottom up through explicitly designed network influence and homophily mechanisms added specificity to VanderWeele's (2011) numerical sensitivity analysis by clarifying the substantive conditions under which Christakis and associates' observational estimates can be regarded as reliable. Zachrison et al.'s (2016: 6, 8) comments on the virtue of this methodological synergy deserves to be fully quoted:

> Beyond these concrete results, our further contribution is to provide a clear, readily extensible framework in which to pragmatically test identification and bias claims in a population in which the underlying data generating mechanism is known with certainty. By comparing the GEE-based estimates against an agent-based model, we improve upon past literature by providing an extensible framework within which possible confounding can be tested. The underlying ABM framework allows detailed, methodologically individual specifications of behavioral interaction rules. These can then be presented to a proposed analytic approach to verify its robustness. (...) An agent-based modeling approach allows rapidly one to generate test populations that may exhibit particular forms of confounding or other threats to identification, and empirically verify the extent to which a regression strategy is or is not susceptible to such a postulated threat.

6.5.2 Network Properties: ABMs and SIENA Models

My second example of how a combination of methods based on different assumptions can increase the credibility of the causal claims these methods are intended to support also concerns the study of the interplay between selection and social influence but, this time, in connection with the effect of these mechanisms on changes in network properties rather than individuals' outcomes. The method at issue here, i.e. stochastic actor-oriented models (SOAM)—also known as "SIENA models" from the general label given the underlying statistical framework (SIENA standing for Statistically Modeling of Longitudinal Network Data [see Snijders 2014])—was designed to describe repeated observations of a given network, and was built with the explicit goal of accounting for both network and behavior changes in such a way that the relative weight of selection and social influence could be identified, given various structural constraints depending on the evolving network structure (see Steglich et al. 2010). Interestingly, Christakis and colleagues themselves recognized that this method may have been exploited to address their causal inference problem but explained that they finally excluded it because of the size of their sample, and also noted that "these models also involve their own assumptions, of course, and these models do not escape some of the general criticisms of the use of observational data, despite any claims to the contrary" (Christakis and Fowler 2013: 567).

Daza and Kreuger (2019: 2) provided a systematic analysis precisely of SIENA's assumptions. For my purpose, the interesting point is that they did so by relying on an ABM, which they explicitly proposed to see, not only as a tool to perform theory exploration and to test mechanism-based explanations, but also as a method "to assess the robustness of statistical methods". To illustrate this perspective, similarly to Zachrison et al.'s (2016) above-mentioned study, Daza and Kreuger (2019: 4, 5) built an ABM to generate network data without missing ties and under perfectly known mechanisms, and then they estimated SIENA models on these data in order to assess under which conditions SIENA can produce reliable and consistent estimates of selection and social influence effects. The intuition of the importance of model diversity in terms of independent assumptions (see Section 6.4) is clearly visible in Daza and Kreuger's (2019: 5) reasoning when, after acknowledging that several sensibility analyses had already been performed on SIENA, namely with respect to its robustness against missing data and different amounts of tie change across waves, they remarked that "(...) those studies use the same data generation mechanisms specified by SIENA. To our knowledge, no published work has assessed the robustness of SIENA estimates using alternative mechanisms." In their view, for the robustness check to be convincing, the alternative generative model should be "different from SIENA but still comparable" (Daza and Kreuger 2019: 9), which essentially means that the alternative mechanisms should primarily remove SIENA simplifications that seem the most overtly unrealistic. If, Daza and Kreuger (2019: 10) argued, SIENA estimates are unaffected by this increased realism, then one may conclude that SIENA existing misspecifications (i.e. simplifications) are tenable for causal identification of selection and social influence effects.

By following this logic, Daza and Kreuger (2019: 11–14) designed an ABM where artificial agents possess one behavioral attribute (a continuous numerical variable defined over 0 and 1) and perform network decisions, i.e. create a tie with a given probability or, if the probability condition is not satisfied, delete a tie. If the tie is created, the agent can choose the most similar agent on the behavioral attribute (homophily mechanism) with a given probability or, if the probability condition is not verified, a randomly selected agent. Agents also possess a social "radius" variable, i.e. a range (defined on a spatial basis) within which they can choose their potential contact. Agents are heterogeneous with respect to the size of this "social circle"—a source of agent-level heterogeneity that is absent from SIENA mechanisms, emphasized Daza and Kreuger (2019: 9, 10, 11, 19). At a certain rate, agents are also influenced by other agents—in particular, the value of a given agent's behavioral attribute becomes closer to that of one of her contacts (randomly selected). Behavioral change at the agent level can happen also independently from social influence (at a rate given by another model's parameter)—a second source of agent-level heterogeneity that is absent from SIENA mechanisms and that may confound both selection and social influence, Daza and Kreuger (2019: 19 and table 2) emphasized.

Given these mechanisms, Daza and Kreuger (2019: tables 1, 4, and 5) simulated the ABM under various combinations of homophily and social influence as well as under various levels of agents' heterogeneity as to the extent of their social "radius" and network-unrelated behavioral changes; then, on the network generated under those varying conditions, they estimated SIENA models with different specifications. They found that, while SIENA was effective in detecting the presence of social influence when the corresponding mechanism was at work within the ABM, SIENA estimates of selection could lead to inference of the presence of this mechanism when in fact it was absent (Daza and Kreuger 2019: 25–26). The authors discovered that, in order to correct this problem, more features of the ABM's generative mechanisms had to be included among the SIENA effects—which was possible in this case because the data-generating process was perfectly known—, and, on the other hand, the sources of heterogeneity that were present in the ABM—namely, variability in the "social radius" and random behavioral change unrelated to social influence—had to be eliminated (Daza and Kreuger 2019: 27, 28). For this reason, the authors finally recommended taking SIENA estimates of the relative importance of selection and social influence with caution when behavioral heterogeneity is likely to be present in the setting under study, and concluded their analysis by noting the virtue of ABMs for reaching this type of scope condition:

> (...) in order to be able to build confidence in our analysis techniques, especially when applied to complex and dynamic systems, it is important to be able to validate them against a diverse set of mechanisms that might plausibly represent real-world situations. With the use of modeling techniques such as ABM, a wide range of previously difficult problems becomes much more accessible. We have conducted a first step in this direction, but much more is to be done to assess complex statistical models like SIENA.

6.5.3   HIV prevalence: ABMs and RCTs

My third research example illustrating the intuition that "evidential pluralism" is increased when the same causal inference problem is addressed at the intersection of methods based on different understanding of causality and mechanisms, thus involving heterogeneous assumptions and, in turn, "independent" simplifications, shifts the focus from statistical methods for observational data to experiments, and concerns the role that ABMs may play as a tool to perform mechanism-based robustness analyses for putative effect-of-a-cause relationships. The piece of research comes from epidemiology and the specific phenomenon under scrutiny is the effect of using pre-exposure prophylaxis (PrEP) on HIV prevalence (see Buchanan et al. 2020).

In particular, Buchanan and colleagues observed that existing RCTs assessing this effect tend to limit the analysis to the potential benefits of PrEP among PrEP users whereas they ignore the potential indirect benefit of PrEP among those who do not take the treatment but have sex with men who do. By ignoring these "disseminated" effects, argued the authors, the estimation of the overall effect of the intervention is potentially biased because the potential positive impact of using PrEP among those who do not take it but are in contact with PrEP users is not assessed. Buchanan et al. (2020: 4) emphasized that this limitation of existing RCTs arises from one of the fundamental assumptions of RCTs that I discussed in Chapter 5 (Section 5.3.1), i.e. the "stable unit treatment value assumption" (SUTVA), which requires that the potential outcome of a given subject is unaffected by the treatment received by another subject (in addition to assuming the absence of hidden variations of the treatment itself). As noted by Buchanan et al. (2020: 17), in the case of the effect of PrEP on HIV prevalence, the lack-of-interference part of the SUTVA assumption is clearly violated by construction because of the underlying (unobserved) sexual risk network in which both users and non-users of PrEP are embedded.

In principle, the issues could be fixed by a two-stage experimental design where groups of individuals who are known to be connected are randomly assigned to a PrEP treatment or control, and then individuals within the treatment group(s) are randomly assigned to receive the treatment or not (see Buchanan et al. 2020: fig. 2). Under this design, and given some assumptions—among which "partial interference" (requiring that interdependence between outcomes does not extend across intervention components)—, the indirect effect of PrEP, i.e. the effect among those who are not treated but are sexually in contact with treated partners, could be quantified by comparing those who were assigned to the intervention group but randomized to not receive the treatment with subjects in a control group. In practice, however, Buchanan et al. (2020: 17) noted that this type of design is difficult to realize—because the underlying sexual network is unknown—and questionable from an ethical point of view. For this reason, they proposed to rely on an ABM with the explicit intent "to emulate a two-stage randomized clinical trial, which may be considered unfeasible or currently unethical to implement in this population" (Buchanan et al. 2020: 5).

In particular, they created an artificial population of 11,245 agents with demographic variables (including race), behaviors (including drug consumption), HIV and clinical status aligned with existing empirical estimates for the population of interest, i.e. men having sex with men aged 18–65 in Atlanta followed over 2 years from 2015 to 2017 (see Buchanan et al. 2020: 5–6). The network of ties between agents representing sexual partnerships was empirically calibrated on the basis of the observed annual number of partners, and so was the number of monthly sexual acts (see Buchanan et al. 2020: 7). To simulate a two-stage randomized trial, at the beginning of the simulation the network components appearing in the artificial sexual network (of size ranging from 2 to 100 agents) were randomized to either PrEP or no PrEP, and, within each intervention component, HIV-negative agents were randomized to PrEP according to the PrEP coverage level (one of the parameters of the simulation). The probability of dyadic transmission of HIV during sexual acts was empirically calibrated, and depended on various factors including exposure to PrEP treatment (see Buchanan et al. 2020: 9).

Within this artificial world, Buchanan et al. (2020: 12) compared the simulated HIV incidence within the intervention and control network components, separately for agents on PrEP and agents not on PrEP, as a function of each component PrEP coverage level. They essentially found that the estimated disseminated effect amounted to an 8% infection risk reduction for agents on no PrEP within treated components compared to agents in the control groups; when the PrEP coverage level increased from 30% to 70%, the estimated disseminated effect increased to a 15% infection risk reduction (Buchanan et al. 2020: 12, 14). The authors were transparent on the limitations of these results—namely, the assumption of a static sexual network, an assumption in turn arising from the necessity of respecting "partial interference" and "stratified interference", two conditions that are required for allowing proper estimates of indirect effects within two-stage experimental designs (Buchanan et al. 2020: 10, 18)—but they also appreciated ABM's potential to quantify biases generated by even more strict assumptions of RCTs typically designed in the field. In their own words:

In this ABM, we observed many scenarios contrasting adjacent coverage levels for which the overall effect estimate was closer to the null, while the composite effect demonstrated a more protective effect, highlighting the importance of considering the suite of disseminated and direct effects when dissemination may be present.

(Buchanan et al. 2020: 17)

6.5.4   HIV treatments: ABMs and DAG-based identification strategies

My last example of method diversity illustrating the intuition that causal inference can be strengthened from combining different methods that rely on assumptions that are likely to involve heterogeneous simplifications, thus independent sources of errors and biases, reverses the direction of the methodology synergy. Until now, I have discussed indeed pieces of research where the ABM was proposed as a tool to submit statistical methods for observational data and experiments to a mechanism-based kind of sensitivity and robustness analysis. What I want to consider finally is a case where a DAG-based identification strategy, namely the door criterion that I discussed in Chapter 5 (see Section 5.3.3.1), was exploited to alert for possible weaknesses in the construction of data-driven ABMs.

In particular, Murray et al. (2017: 132) were interested in the effect of offering HIV-positive individuals antiretroviral therapy on their 12-month mortality risk. By assuming that the treatment can be specified as "always treat" and "never treat", and in the case where an RCT was unfeasible, they proposed to compare the estimate of this effect generated by an ABM where agents are assigned to the treatment according to these two strategies with the estimate produced by the parametric g-formula (for a description of this method, see Hérnan and Robins 2020: ch. 13).

In the case of an ABM, Murray and colleagues assumed that each agent possesses a specific value (at a given point in time, say, the month k) of the variable representing her level of white blood cells—CD4 cell count, hereafter—as well as a given state on the indicator for death at the beginning of the period considered (say, the month); the probability of initiating the treatment at a given point in time is a function of CD4 cell count history and the probability of dying depends on both the CD4 cell count and treatment history. Murray et al. (2017: 133) emphasized that both conditional probabilities can be modeled with different functional specifications and that the values for the parameters of the chosen function could be based on published estimates obtained through observational and/or experimental studies. On this basis, Murray and colleagues continued, simulated individuals' trajectories could then be obtained by setting to 1 and to 0, alternatively, the treatment probability for all times and conduct Monte Carlo simulations on a large population of agents. The comparison of the 12-month mortality risk for the two series of simulations would give an estimate of the effect of the treatment. When using the parametric g-formula, all variables and functional specifications of the conditional probabilities would be the same; the only difference is that the value of the parameters of these specifications come from a specific study, namely "a follow-up study of HIV-positive individuals with monthly measurements of CD4 cell count, treatment, and mortality" (Murray et al. 2017: 133).

Now, Murray et al.'s (2017: fig. 2) essential point is that a DAG-based representation of the setting possibly implemented in both the ABM and the g-formula allows one to see clearly that this setting involves a "treatment-confounder" feedback arising from the fact that the CD4 cell at time k is affected by treatment at $k - 1$ and affects treatment at time k; moreover, if an unmeasured factor exists that affects both the CD4 cell and the mortality outcome, then the CD4 cell becomes a collider, which makes this variable unusable for adjustment (a problem that I discussed in Chapter 5, Sections 5.3.2–5.3.3). The consequence of this is that the parameter quantifying the conditional association between the treatment at $k - 1$ and mortality within the models specifying the conditional probability of mortality may be different from zero even though the causal direct effect associated with the treatment is nil. As long as an ABM relies on this potentially biased parameter for each agent within the artificial population, Murray et al. (2017: 133) argued, the ABM can lead to estimates of the average treatment effect that are severely biased, a phenomenon that they demonstrated by running the ABM under various distributions of potential unmeasured confounders affecting both the CD4 cell and the mortality outcome (Murray et al. 2017: table 1).

Although one may remark that Murray et al.'s ABM in fact essentially is a micro-simulation model (on this point, see also Arnold et al. 2019: 2), their analysis conveys the important message that ABMs that are based on empirically calibrated parametric models for the agent's behavior should pay particular attention to the way the model at hand was estimated in order to avoid injecting biases within the simulation. This clearly is an area where DAG-informed modeling and ABMs can fruitfully communicate.

To conclude, let me note that, while the four pieces of research that I have discussed clearly proposed different types of synergies between experiments, statistical methods for observational data, and ABMs for causal inference, they share a common point: the authors of these articles had the impression that explicit comparisons of different methods on the same causal issue are extremely rare, and that their work represented an exception. In particular, Daza and Kreuger (2019: 2), Buchanan et al. (2020: 18), and Murray et al. (2017: 132) agreed on the fact that these methods continue to be used in isolation, that they receive unequal coverage in university training (ABMs being less extensively and systematically covered), and that, because of such lack of training, many scholars continue to ignore their similarities and differences (see also Arnold et al. 2019). In practice, this state of affairs increases the probability that the mutual and synergetic contribution to causal inference of ABMs and experimental/observation methods continue to be ignored or, at least, happily downplayed (see also Halloran et al. 2017). The goal of the pluralistic view on causality and the evidential pluralism perspective that I attempted to defend in this chapter aims at contributing to reverse this trend.

Coda

Despite the rapid diffusion of agent-based models (ABMs) within a large spectrum of disciplines, a systematic analysis of the possible contribution of ABMs to causal inference is still lacking. In a brief article discussing four possible candidate accounts of causal explanations for ABM—namely, "agent", "algorithmic", "intervention", and "mechanism" causation—, Anzola (2020: 48, 51) overtly observed:

> In agent-based computational social science (ABCSS), however, causation has been systematically neglected in the discussion about social explanation using agent-based modelling. Most references to the concept in the field are very general, e.g. through the notions of 'causal mechanism', 'causal relations' or 'causal structures', or are linked to abstract conceptual issues, e.g. downward causation. (...) References to causation in ABCSS are scarce and usually associated with basic intuitions or platitudes about causation.

This book wanted to fill this gap by connecting systematically two lines of inquiry. On the one hand, I examined various understandings of the concepts of causation and mechanisms, and described how they are at work within ABMs and various identification strategies of causal effects that are widely accepted as proper tools for causal inference. On the other hand, I systematically compared ABMs with typical illustrations of these strategies, in particular randomized control trials (RCTs), instrumental variables (IVs), and various forms of directed acyclic graph (DAG)-based approaches, in order to establish under which conditions ABMs and experimental/observational methods can support persuasive causal claims in practice. In the past, similar efforts of conceptual and methodological clarification have been produced with respect to other research traditions that were gaining momentum. For instance, when small-N research reached a critical mass, some felt it necessary to investigate in what sense small-N studies allow causal inference (see Mahoney 2000).

With this ambition in mind, first, I approached ABMs, experiments, and statistical methods for observational methods from the point of view of the intuitions about causality and mechanisms which animate them: a production view of causality and a vertical view of mechanisms for ABMs; a dependence view of causality and a horizontal view of mechanisms for experimental and observational methods. On this conceptual basis, I then asked under which conditions these two classes of methods can produce persuasive data and arguments supporting causal claims along the lines of the specific understanding of causality embedded within their internal machinery.

My conclusion was that, although both ABMs and experimental/observational methods are *in principle* self-sufficient to produce knowledge that is relevant for causal inference in their own terms, *in practice* both families of methods face challenges concerning data limitation and testability of assumptions. Both approaches make, necessarily, use of formal and substantive assumptions, not all of which are empirically defensible. Both approaches have at their disposal reliability tests, which, however, are partly based on conventions and incapable of conclusively ruling out errors. In both cases, the data and arguments produced are likely to be imperfect.

As a consequence, causal claims from ABMs, experiments, and observational methods ultimately equally require a complex mix of data, assumptions—partly based on background, subject-specific, and expert arguments, which necessarily come from outside the method and the data under scrutiny—, and reliability checks that are necessary to assess the sensitivity and the robustness of results to assumptions only partially supported by data. Thus, as I did not find in the literature knock-down arguments which would prove the superiority of either the ABM approach to causal inference or experimental/observational methods of causal inference, I finally argued in favor of a combination of the two classes of methods so that the diverse type of data and arguments that they are capable of producing can compensate their respective limitations.

I see three main potential objections to this line of reasoning. I discuss each of them in the following section (Section 1) before providing a concluding summary of my analysis (Section 2).[1]

1   Possible Objections

At various points of my investigation, I have discussed several possible objections to specific aspects of my argument. In particular, I examined the view according to which the ABM's distinctiveness as to its capacity to study mechanisms should not be over-emphasized compared to observational methods because both approaches ultimately are made of "variables" (see Chapter 2, Section 2.6). I also addressed the claim that, as ABMs always depend on data that are exogenous to them, they could not produce "evidence" *on their own* by construction (see Chapter 4, Section 4.1.5). Finally, I discussed the theses of those who believe that the "superiority" of experimental and observational methods for causal inference ultimately rely on their "formal" assumptions, their "materiality", and their reliability procedures, all features on which ABMs would be deficient (see Chapter 5, Sections 5.4.1–5.4.3).

I would like to focus now on possible objections to my argument that are more general in that, if they hold, they would undermine the entire infrastructure of the analysis rather than this or that point.

---

[1] These sections build on Casini and Manzo (2016: 62–9).

1.1 Causation is Not Constitution

Among these objections, the first that deserves to be discussed concerns the legitimacy of the question from which this book started, i.e. the sense in which, if any, ABMs can be relevant for causal inference. I suspect indeed that many quantitative social scientists may regard this question as being itself neither here nor there.

From a philosophical point of view, this radical objection would arise from the widespread view that causation and constitution are different kinds of relations (see Craver and Bechtel 2007). Whilst causation obtains between spatiotemporally distinct entities (cause and effect are distinct events or states) and is asymmetric (intuitively, effects depend on causes), constitution obtains between spatiotemporally overlapping entities (parts and wholes) and is non-asymmetric (in a qualified sense, parts depend on wholes and *vice versa*). Given this distinction, if, as I argued, ABMs study how changes in dynamic loops between micro- and macro-level variables trigger changes in other macro-level variables, then ABMs study constitutional rather than causal relations. Consequently, ABMs are in the business of generative explanation, and not of causal inference.

From within sociology, one may make a parallel point, by arguing that I ignored the distinction between "description" and "explanation". As restated by Goldthorpe (2016: ch. 1), description concerns detecting, or making "visible", probabilistic regularities in the social world, whereas explanation concerns providing narratives, which make "transparent" how these regularities were brought about. One thing is to see regularities. To this end, experiments and multivariate statistics are very good. Another thing is to see through them. To this end, theory is needed. With this distinction in mind, one may conclude that, if ABMs, as I argued, aim to establish which low-level chains of events lead to observed robust connections between aggregates, then ABMs are not a method for causal inference but a device for generating (potential) explanations. In other words, so the objection goes, while/if ABMs are concerned with explaining causal states of affairs on the assumption that they obtain, a proper method for causal inference is concerned with establishing whether causal states of affairs obtain in the first place.

My reaction to this objection is that it stems from a dependence account of causation, and ignores what causality may mean from a production point of view. One point that I wanted to make was precisely that several understandings of causation are equally legitimate and that the data and arguments produced from within the methods animated by these various views are equally necessary to reduce the uncertainty about the non-accidental nature of a given observed dependence among happenings. From this point of view, both experimental/observational methods and ABMs are methods for causal inference.

Within experimental/observational methods, causal inference is understood as generalizing correlations from the sample under scrutiny to a target population. This perspective is tied to a dependence/difference-making view of causality. Description is an activity that emanates from this view. Within ABMs, causal inference is conceived of as extrapolating the mechanism connecting different levels (or scale) of analysis neatly described *in silico* to the real world. This perspective on causal inference, in contrast, is tied to a production view of causality. Explanation is an activity associated with this understanding of causality.

The ultimate aim of my argument is to stop trying to establish a hierarchy between intuitions about causation, and between descriptive and explanatory activities. My argument is based on the conviction that the internal structure of different methods makes them more appropriate to pursue different sorts of causal analysis. However, this difference should be understood in collaborative, and not conflicting, terms, insofar as different kinds of data and argument are necessary to establish causal claims. The view opposing causation (description) and constitution (explanation), in contrast, contributes to fueling the "causal exclusivism" that, as I argued in Chapter 6, should be abandoned in favor of an "evidential pluralism" perspective. From a philosophical point of view, the concept of "development explanations", i.e. a type of explanation that combines causal and constitutive dependences (see Ylikoski 2013: 292–4), clearly points in that direction and can help the methodological synergy I am advocating here.

1.2   Lack of a Specific Research Strategy

A second objection that one may address to the proposed synergy between ABMs, experiments, and statistical methods for observational data is that the division of labor implied by this synergy was not clearly spelled out on a methodological level.

From a philosophical point of view, this objection would arise from the observation that the "evidential pluralism" perspective that I defended did not translate into a specific recipe for combining different types of data and arguments. This would reveal, philosophers may argue, a more fundamental problem: the pragmatic view of evidence and the "coherentist" approach to epistemic justification that I exploited in Chapter 6 are insufficient to ground my argument on a fully developed theory of evidence. While I appreciate the importance of this objection in theory, the two following elements should be kept in mind fully to appreciate how the objection is consequential in practice.

First, philosophers themselves are in disagreement about the utility of available philosophical theories of evidence for empirically minded scientists (Achinstein 2000; Cartwright et al. 2010); moreover, among philosophers, the problem of how integrating different types of evidence continues to be an open one (Williamson 2015).

Second, from a sociological point of view, my analysis in fact implies a clear *modus operandi* for causal reasoning in social sciences. According to this *modus operandi*, establishing persuasive causal claims would ideally amount to looping the following steps. First, one should employ experimental and/or statistical methods for observational data to show that, given the best data available, the probability that the connection between a set of happenings of interest is observed is higher than by chance alone. This step is performed from within a dependence (or difference-making) account of causality. Second, one should formulate a clear set of hypotheses about the underlying chains of actions, interactions, and their constraints, which are believed to be responsible for the observed connection. This step is performed from within a production understanding of causality and a vertical view of mechanisms. Third, the hypotheses should be translated into a formal model, whose behavior should be simulated in order to see if the dynamics triggered by the model are able to generate the observed connection. For this step, I have argued that ABMs are an especially powerful tool because of the granularity they allow for theoretical mechanism design. Fourth, the simulation should be constrained by as much empirical information as possible on the input side (i.e. empirical calibration) in order to see if, when as many pieces of the simulation machinery as possible are based on empirical regularities, the ABM's dynamics are still able to generate the observed connection of interest. Finally, formal and substantive assumptions for which empirical data are lacking or sufficient should be submitted to systematic robustness and sensitivity analysis in order to assess the extent to which the ABM's results are stable across variations of those assumptions.

In this sense, contrary to the objection, my plea for "evidential pluralism" in fact leads to a specific research path in which dependence (or difference-making) and production accounts of causality (as well as horizontal and vertical views of mechanism) interpenetrate. This is especially visible in the third and fourth steps of the research strategy just described: the empirical validation and calibration of an ABM indeed requires information obtained from experimental and observational methods.

It seems to me that this research path is consistent with existing proposals from empirically minded scholars. In particular, Goldthorpe's (2016) manifesto for sociology as a population science defends a research program that combines multivariate statistical techniques, which generate sophisticated descriptions of macrolevel probabilistic regularities, with theoretical reasoning, which offers narratives at the level of actions and interactions that explain the robust patterns detected. Compared to Goldthorpe's proposal, my own agenda puts greater emphasis on an ABM as a tool to provide not only proofs of the "generative" sufficiency of a set of hypotheses but also to produce, through empirical calibration, data and arguments on the empirical realism of the simulated narrative about the connections across levels (or scales) of analysis that the narrative describes. In this respect, the ideal sequence of research steps implied by my plea for "evidential pluralism" amounts to an approach to causal reasoning that is very much in line with some variants of the sociological perspective known as analytical sociology (for an overview, see Manzo 2014a, 2021).

1.3 A Limited Methodological Spectrum

This leads to the last potential objection that I would like to discuss. In short, this objection would attack my investigation for its limited scope. Although for different reasons, both quantitative and qualitative scholars may formulate this critique.

Among quantitative scholars, experimentalists and users of multivariate statistics may question the subset of methods for causal inference that I selected to defend the thesis that ABMs and experimental/observational methods are in fact equally incapable of building causal claims by relying only on the empirical data under scrutiny. To this, my reply would be that, although RCTs, IVs, and DAG-based modeling cannot do full justice to the variety of methods for causal inference, the selection covers the basic types of identification strategies of causal effects, and thus this subset of methods seems sufficient to illustrate the general point of interest: no matter what specific tool is chosen, design- and model-based identification strategies of causal effect, too, repeatedly rely on assumptions that, because of data limitation and/or logical construction, cannot be tested empirically. The fractal nature of this problem, meaning that for testing some assumptions other assumptions are needed, is effectively illustrated by Berk (2010: 482) when, commenting on regression, he states: "For both the diagnostics and the remedies, new and untestable assumptions are required even before one gets to a number of thorny technical complications".

As to the smaller community of mathematical sociologists who, in addition to experimental and statistical methods, routinely work with analytically tractable mathematical models (like differential equations or game-theoretical models), they may complain about my exclusive focus on a specific form of computational modeling, namely ABMs. To this, I would reply that my admittedly restricted focus seems justified given that the ABM has only recently entered the sociological toolbox, which explains, I believe, why a principled and systematic assessment of its relevance for causal reasoning is still missing. In addition, I explained the deep technical reasons that give ABMs a strong capacity to relax unrealistic assumptions, which leads to higher granularity in mechanism design, as well as a great deal of flexibility in handling the dynamic process of moving from lower to higher levels of analysis compared to other forms of mathematical and computational modeling (see Macy and Flache 2009). For this reason, among methods that are based on a production view of causality, the ABM seems a more flexible method for causal inference, which justifies my choice of comparing it with more traditional observational methods of causal inference.

Finally, more qualitatively oriented scholars may also regard my investigation as having a limited generality, on the grounds that my argument on the relevance of ABMs for causal inference is silent about the role that qualitative evidence may play in establishing persuasive causal claims. The focus on quantitative methods arose from the fact that the primary target I had in mind when starting to think about this book were statistically minded sociologists who, often off the record, deny that ABMs have any value for causal reasoning. Thus, the focus on experiments and statistical methods for observational data should not be seen as a sign that combining qualitative and quantitative information is not part of the "evidential pluralism" perspective that I ultimately advocated.

Quite the contrary, I believe that qualitative data, namely from ethnographic observations, in-depth interviews, document or historical archives, can play a fundamental role, in particular at the stage of empirical calibration of an ABM, thereby helping to adjudicate between different low-level assumptions that may, within the virtual world of the simulation, lead to similar results. In other words, qualitative evidence may help to reduce the problem of "equifinality" (or "multiple realizability"), thereby increasing the credibility of claims about the external validity of the simulated mechanisms. That is why a recent stream of research in the literature on ABMs focuses on how to inject qualitative evidence within these models (see Edmonds 2015; Ghorbani et al. 2015), and pieces of research exist where field data of actors' reasons, choices, and social networks are exploited at the same time to calibrate the ABM and to understand its internal dynamic (for an example, see Manzo et al. 2018).

On the other hand, it is also meaningful to remark that prominent qualitative scholars, too, in particular from within the "process tracing" approach (for a philosophical discussion, from the point of view of causal inference, see Steel 2007: ch. 8), acknowledged that simulating a qualitatively inspired narrative through ABMs can help "to check the plausibility of inferences about causal mechanisms derived from process tracing" (Bennett and Checkel 2014: ch. 1). These scholars, too, defended the view that "diversity and independent evidence are useful in testing explanation".

Thus, although I had a quantitative target in mind and although I focused on a specific simulation-based technique, my argument to the point that methodological synergy and evidential pluralism help to build persuasive causal claims in fact aligns with arguments put forward by well-developed and identified qualitative approaches.

2 Summary

Sociologists have often stressed the importance of developing an autonomous reflection on causal analysis for sociology. For instance, while commenting on Liberson's skepticism against randomized experiments as the gold standard for sociological causal analysis, Goldthorpe (2001: 8) claimed:

> (...) sociologists have to find their own ways of thinking about causation, proper to the kinds of research that they can realistically carry out and the problems that they can realistically address.

By relying on selected works from sociology, economics, epidemiology, computer science, and philosophy of the social sciences, this book aimed at contributing to this meta-theoretical project about causal reasoning in quantitative sociology. To this end, I chose a specific methodological entry. The question that guided this book concerned whether a specific simulation technique, namely the ABM, can produce data and arguments that are relevant to establish causal claims. In particular, I investigated *in what sense* and *under which conditions*, if any, this is possible. The possible value of ABMs for causal inference seemed to me a relevant question for three main reasons.

First, while the ABM is entering the accepted toolbox of quantitative sociologists, markedly different views are still expressed as to its capacity to contribute to empirical research. I documented this disagreement in the Introduction. As a consequence, a principled and systematic assessment of its causal value seemed to me necessary. Second, given the link between ABMs and mechanism-based explanations, the issue of the causal value of this technique allows one to connect scholarship on causation, social mechanisms, and simulation methods that are rarely considered at once. Finally, since assessing the possible value of ABMs for causal inference requires that alternative methodological approaches are also considered, a systematic, and explicit, comparison between ABMs, experiments, and observational methods becomes possible in relation to the specific task of causal inference. To the best of my knowledge, a comparison of this kind is still missing.

Upon endorsing this comprehensive perspective, I first inquired the sources of the observed disagreement about the possible contribution of ABMs to causal inference. I identified two of them.

In Chapter 1, I first documented the extent to which claims on the alleged impossibility of ABMs to contribute to causal analysis compared to experimental and observational methods rely on specific views on what causality and mechanisms are. In particular, building on types of theories of causation identified by philosophers, I showed that dependence accounts of causality—according to which causes make a non-spurious difference to their effects—are contrasted to production accounts—according to which causes bring about their effects via a well-detailed chain of events. In addition, I documented that this opposition is reinforced by the fact that contrasting views on causality square with two opposing understandings of mechanisms, with the followers of dependence accounts of causality regarding a mechanism as a chain of intervening variables (I have labeled this perspective the "horizontal view" of mechanisms) and the followers of production accounts regarding a mechanism as a set of entities and activities that dynamically produce an outcome through iterative aggregation steps across different levels (or scales) of analysis (I have labeled this perspective the "vertical view"). My final diagnosis was that scholars who endorse the former view tend to be skeptical about the potential of ABMs for causal analysis whereas those who accept the idea of "generative causality" tend to see ABMs as a crucial tool for establishing causal claims.

After having explained in Chapter 2 the technical reasons that make an ABM a flexible tool to implement a "vertical" view of mechanism, thus offering a methodological support for the "production" account of causality, in Chapter 3 I documented the second reason for the difference in appreciations of the possible contribution of ABMs to causal inference. This reason is the variety of types of ABMs that populate the literature across several disciplines. Among these types, speculative ABMs are still by far more frequent than theoretically plausible (meaning solidly grounded in existing sociological and psychological theories at the micro-level), empirically calibrated, and validated ones. This led many observers to take a pattern of current applications for an intrinsic limitation of the technique, and to judge ABMs as fundamentally incapable of integrating appropriate empirical information. As a consequence, even among scholars who overtly endorse a production understanding of causation, the ABM is seen as promising for exploring the "causal adequacy" of a mechanism but not to prove that it is actually at work. As put by Goldthorpe (2016: ch. 9), "(...) to show the generative sufficiency of a mechanism is not to show that it is in fact this mechanism that is in some particular instance at work".

Having accepted this variety of views on causality and mechanism and having acknowledged the full range of technical options offered by ABMs in terms of communication between an ABM and external sources of data, I then entered the second part of my study by trying to identify under which conditions this method can contribute to causal inference.

In particular, I argued in Chapter 4 that an ABM can produce knowledge that is relevant for causal inference (from a production view of causality and a vertical understanding of mechanism) when it acts as a "mimicking device" (according to a metaphor by Mary Morgan), i.e. when it generates the high-level dependence relationship of interest (meaning that empirical validation is performed) under assumptions that were in turn based on empirical data proving their realism (meaning that empirical calibration is also performed). More particularly, in Section 4.1.5, I clarified that, when these conditions are met, an ABM can produce causally-relevant knowledge on the connections between the model's low-level components that are empirically calibrated, and the macroscopic patterns of interest. In other words, the proper contribution of an empirically calibrated ABM to causal inference concerns the connection across levels (or scales) of analysis in the real world, and not the low-level mechanisms employed to create these connections—the empirical evidence for these mechanisms in fact comes from outside the ABM through empirical calibration; thus the ABM cannot be considered as causally-relevant *on its own* with regard to these "input" mechanisms.

To the potential objection that the identified conditions under which an ABM can contribute to causal inference are difficult to fulfill in practice because of the chronic lack of sufficiently fine-grained data, I replied that ABMs can rely on various reliability tools (namely, sensitivity, robustness, and dispersion analysis) to quantify the extent to which the ABM's outcomes depend on those pieces of the model for which empirical information is partial or missing. On the other hand, several, although not yet formalized, strategies exist to describe the dynamic through which the model generated the outcome of interest (this is what I called "model analysis"). Both the quantification of the outcome's stability and the transparency of the model's dynamics, I argued, are key to increasing an observer's confidence in the external validity of the across-level consequences of the low-level mechanism postulated by the ABM.

As testified by several quotations that I reported in the book's Introduction, scholars animated by a dependence account of causality (and a horizontal view of mechanisms) would counter-object to the thesis defended in Chapter 4 that, as long as some pieces of the internal machinery of an ABM only rely on reliability checks, and not on data, the model's capacity to tell us whether the postulated mechanism actually works in the real world—and thus its capacity to produce knowledge of a production kind that is relevant for causal inference—remains very limited.

In order to show that the ABM is not exceptional with respect to the fact that its contribution to causal inference ultimately relies on a complex mix of data and arguments, I then pushed further my analysis in Chapter 5. There I scrutinized methods that rely on information collected by letting individuals act in real systems, be these systems experimentally created or "naturally" given; these methods are usually regarded as good methods for causal inference (from the point of view of dependence or difference-making accounts of causation). In particular, I focused on RCTs, IVs, and various forms of DAG-based modeling, and I showed that these methods, too, ultimately justify their causal conclusions by arguments that do not exclusively rely on empirical data. The simple reason for this is that, similarly to ABMs, the assumptions required to causally interpret the observed connections cannot be tested empirically because of lack of data or because it is impossible by construction to empirically adjudicate the truth of the assumptions.

Thus, I concluded, ABMs, experiments, and statistical methods for observational data can all reach their respective goals, namely convincing an observer of, respectively, the external validity of the postulated chains of events and the non-spurious character of the detected dependence, only by combining partial empirical information with theoretical, subject-specific knowledge that is exogenous to the data directly used to calibrate/validate the model or document the putative connection.

In Section 5.1, I acknowledged that several statisticians and sociologists have already forcefully argued that the causal interpretation of statistical estimates requires more than empirical data alone (for a review of this debate, see Goldthorpe 2016: ch. 8). In this sense, my argument is not new. However, I believe that the original and, I suspect, more controversial part of my analysis is the implication that I drew from this argument, namely that, upon scrutiny, there is no cogent argument to the point that experimental and observational methods are *superior* to other methods, namely simulation-based methods, in constructing persuasive causal claims.

In economics, by partly building on the work of Nancy Cartwright, which I also considered, Angus Deaton has recently defended this argument with respect to randomized experiments. Based on an analysis of the "ideal conditions" and the "practical problems" of implementation of this method, Deaton (2010: 426) claimed:

> Randomized controlled trials cannot automatically trump other evidence, they do not occupy any special place in some hierarchy of evidence, nor does it make sense to refer to them as "hard" while other methods are "soft". These rhetorical devices are just that; metaphor is not argument, nor does endless repetition make it so.

Essentially, my analysis has consisted in extending this view to statistical methods for observational data and computational models, namely ABMs.

Over and above the fact that all methods for causal inference ultimately rely on a complex mix of data and arguments—such that, when properly understood, any of them can be claimed to be more "empirical" than others—there is a deeper reason that motivates my rejection of the view that one class of methods—and the specific understanding of causality and mechanisms behind them—would be more fundamental, and thus more appropriate, to defend persuasive causal claims.

This reason, which I discuss in Chapter 6, is that the knowledge produced by each family of methods, no matter how firmly grounded in theory and data, can only partly capture the information needed to reduce the uncertainty about the causal nature of a given dependence relationship.

On the one hand, data and arguments on a credible sequence of events sustained by interactions among (credible) entities and activities (i.e. a mechanism in a vertical sense) influences one's belief that some relation is genuinely causal. Intuitively, empirically observed robust dependences not backed by knowledge on the entities and activities responsible for them do not eliminate the doubt that the relation is affected by some sort of systematic bias. On the other hand, data and arguments on a robust dependence, too (i.e. a mechanism in a horizontal sense), influence one's belief that the relation is causal. Intuitively, in fact, data and arguments of how entities interact to produce some behavior does not guarantee that their organized operation is not masked in the broader context of further (possibly unknown) modes of interaction and outer influences. That is why, I finally argued, mechanistic knowledge (in the vertical sense) should not be seen as more fundamental than difference-making knowledge, or *vice versa*. Rather, the two kinds of data and arguments complement each other. This reasoning has ultimately motivated my plea for "evidential pluralism" and for a methodological synergy between ABMs, experiments, and statistical methods for observational methods, which only allows us to generate different but complementary kinds of knowledge, equally necessary to establish persuasive causal claims.

Thus, with this book, my hope is to contribute to shifting the focus of current methodological debates from defending the superiority of particular methods, through the often implicit endorsement of particular notions of causality and mechanism, to discussing how different methods, hence a different understanding of causation and mechanisms, can be combined to produce different kinds of data and arguments that can complement each other's weaknesses. Goldthorpe (2016: ch. 9, emphasis added) seemed to go in the same direction when, with regard to the specific task of proving the empirical adequacy of a postulated mechanism (in my terminology, a vertical mechanism), he claimed:

> What is important is that the actual operation of mechanisms should be tested in *as many ways as is possible* and the *results obtained be considered in relation to each other*. It should not be expected that any particular test will produce 'clinching results', at least not of a positive kind, but at best 'vouching' results—to take up Cartwright's (2007b: ch. 3) useful distinction; and greatest weight has then to be given to *how far results from different tests do, or do not, 'fit together'*.

By building on a principled and systematic assessment of the possible contribution of a specific simulation technique, namely the ABM, to causal inference, and comparing it to experimental and observational methods, this book attempted to generalize this synergistic view to causal reasoning as a whole. In Chapter 6 (Section 6.5), I commented on several pieces of research that illustrated this project by focusing on the same causal problem and challenging each other's results through different methodologies. In particular, ABMs were exploited as a tool to perform mechanism-based robustness analysis on putative causal effects generated by experiments and statistical methods for observational data. However, those works were realized separately by different teams with different methodological skills. For now it is clearly difficult to find a single piece of work by the same scholar(s) that implements the entire methodological synergy I am advocating here. I like to hope that the meta-theoretical but at the same time methodologically informed analysis proposed in this book will succeed in fostering and informing further investigations along this synergistic approach to causal inference.

References

Abbott A. (1988). Transcending General Linear Reality. *Sociological Theory*, 6:169–186.

Abbott A. (1997). Seven Types of Ambiguity. *Theory and Society*, 26:357–391.

Abbott A. (1998). The Causal Devolution. *Sociological Methods and Research*, 27(2):148–181.

Abbott A. (2001). *Time Matters: On Theory and Method*. Chicago: University of Chicago Press.

Abbott A. (2007). Mechanisms and Relations. *Sociologica*, 2:1–22.

Abend G., Petre C., and Sauder M. (2013). Styles of Causal Thought: An Empirical Investigation. *American Journal of Sociology*, 119:602–654.

Achinstein P. (2000). Why Philosophical Theories of Evidence are (and Ought to Be) Ignored by Scientists. *Philosophy of Science*, 67:S180–S192.

Ajelli M., Gonalves B., Balcan D., Colizza V., Hu H., Ramasco J. J., Merler S., and Vespignani A. (2010). Comparing Large-Scale Computational Approaches to Epidemic Modeling: Agent-based versus Structured Meta-population Models. *BMC Infectious Diseases*, 10(190):1–13.

Ajelli M., Merler S., Pugliese A., and Rizzo C. (2011). Model Predictions and Evaluation of Possible Control Strategies for the 2009 A/H1N1v Influenza Pandemic in Italy. *Epidemiology & Infection*, 139(1):68–79.

Aksoy O. and Gambetta D. (2016). Behind the Veil: The Strategic Use of Religious Garb. *European Sociological Review*, 32(6):792–806.

Alexander J. M. (2007). *The Structural Evolution of Morality*. Cambridge: Cambridge University Press.

Amati V. and Stadtfeld C. (2021). Network Mechanisms and Network Models. (Ch. 23) In Manzo G. (ed.), *Research Handbook on Analytical Sociology*, Ch. 23. Cheltenham (UK): Edward Elgar.

Andersen O. (2014a). A Field Guide to Mechanisms: Part I. *Philosophy Compass*, 9(4):274–283. 10.1111/phc3.12119.

Andersen O. (2014b). A Field Guide to Mechanisms: Part II. *Philosophy Compass*, 9(4):284–293. 10.1111/phc3.12118.

Anderson L. R., Bukodi E., and Monden C. W. S. (2021). Double Trouble: Does Job Loss Lead to Union Dissolution and Vice Versa? *European Sociological Review*, https://doi.org/10.1093/esr/jcaa060.

Angrist J. D. and Krueger A. B. (1991). Does Compulsory School Attendance Affect Schooling and Earnings? *Quarterly Journal of Economics*, 106(4):979–1014.

Agent-based Models and Causal Inference, First Edition. Gianluca Manzo.
© 2022 John Wiley & Sons, Inc. Published 2022 by John Wiley & Sons, Inc.

Angrist J. D. and Krueger A. B. (2001). Instrumental Variables and the Search for Identification: From Supply and Demand to Natural Experiments. *Journal of Economic Perspectives*, 15(4):69–85.

Angrist J. D. and Pischke J. (2010). The Credibility Revolution in Empirical Economics: How Better Research Design is Taking the Con Out of Econometrics. *Journal of Economic Perspectives*, 24:3–30.

Antonakis J., Bendahan S., Jacquart P., and Lalive R. (2010). On Making Causal Claims: A Review and Recommendations. *The Leadership Quarterly*, 21:1086–1120.

Anzola D. (2020). Causation in Agent-Based Computational Social Science. (Ch. 5) In Verhagen H., Borit M., Bravo G., and Wijermans N. (eds.), *Advances in Social Simulation: Looking in the mirror*, Springer Proceedings in Complexity, pages 47–62. Springer Nature Switzerland AG.

Archer M. (1998). Introduction: Realism in the Social Sciences. (Ch. 7) In Archer M., Baskar R., Collier A., Lawson T., and Norrie A. (eds.), *Critical Realism: Essential Readings*. London: Routledge.

Arnold K. F., Harrison W. J., Heppenstall A. J., and Gilthorpe M. S. (2019). DAG-informed Regression Modelling, Agent-based Modelling and Microsimulation Modelling: A Critical Comparison of Methods for Causal Inference. *International Journal of Epidemiology*, 48(1):243–253.

Arthur W. B. (2006). Out-of-Equilibrium Economics and Agent-Based Modeling. In Tesfatsion L. and Judd K. L. (eds.), *Handbook of Computational Economics. Agent-based Computational Economics*, Vol. 2, pages 1551–1564. Amsterdam: North Holland Elsevier.

Arthur W. B. (2021). Foundations of Complexity Economics. *Nature Reviews Physics*, 3:136–145. https://doi.org/10.1038/s42254-020-00273-3.

Arthur W. B., LeBaron B., Palmer B., and Taylor R. (1997). Asset Pricing under Endogenous Expectations in an Artificial Stock Market. In Arthur W. B., Durlauf S. N., and Lane D. A. (eds.), *Economy as an Evolving Complex System II*, Vol. XXVII, pages 15–44. Santa Fe Institute Studies in the Science of Complexity, Reading, MA: Addison-Wesley.

Auchincloss A. H. and Roux A. V. D. (2008). A New Tool for Epidemiology: The Usefulness of Dynamic-Agent Models in Understanding Place Effects on Health. *American Journal of Epidemiology*, 168(1):1–8.

Aven B. L. and Hillmann HillmanHi H. (2017). Structural Role Complementarity in Entrepreneurial Teams. *Management Science*, 64(advance online publication):https://doi.org/10.1287/mnsc.2017.2874).

Axelrod R. (1997). *The Complexity of Cooperation: Agent-Based Models of Competition and Collaboration*. Princeton: Princeton University Press.

Axtell R. L. (2000). Why Agents? On the Varied Motivations for Agent Computing in the Social Sciences. Technical report, Center on Social and Economic Dynamics, Brookings Institution, Washington, DC. Working Paper 17.

Axtell R. L. (2001). Effects of Interaction Topology and Activation Regime in Several Multi-Agent Systems. In Moss S. and Davidsson P. (eds.), *Multi-Agent-Based Simulation: Lecture Notes in Computer Science*, pages 33–48. Berlin: Springer.

Axtell R. L. and Epstein J. M. (1996). *Growing Artificial Societies: Social Science from the Bottom Up*. Washington, DC: Brookings Institution Press.

Axtell R. L., Epstein J. M., Dean J. S., Gumerman G. J., Swedlund A. C., Harburger J., Chakravartya S., Hammonda R., Parkera J., and Parkera M. (2002). Population Growth and Collapse in a Multi-agent Model of the Kayenta Anasazi in Long House Valley. *Proceedings of the National Academy of Sciences*, 99:7275–7279.

Axtell R. L., Epstein J. M., and Young P. H. (2006). The Emergence of Classes in a Multi-agent Bargaining Model. In Epstein J. M. (ed.), *Generative Social Science: Studies in Agent-based Computational Modeling*. Princeton: Princeton University Press.

Balke T. and Gilbert N. (2014). How Do Agents Make Decisions? A Survey. *Journal of Artificial Societies and Social Simulation*, 17(4):13. http://jasss.soc.surrey.ac.uk/17/4/13.html.

Barberousse A., Franceschelli S., and Imbert C. (2009). Computer Simulations as Experiments. *Synthese*, 169:557–574.

Barringer S. N., Eliason S. R., and Leahey E. (2013). A History of Causal Analysis in the Social Sciences. In Morgan S. L. (ed.), *Handbook of Causal Analysis for Social Research*, pages 9–26. Dordrecht: Springer.

Baumgartner M., Casini L., and Krickel B. (2020). Horizontal Surgicality and Mechanistic Constitution. *Erkenntnis*, 85:417–430. https://doi.org/10.1007/s10670-018-0033-5.

Becker J., Brackbill D., and Centola D. (2017). Network Dynamics of the Wisdom of Crowds. *Proceedings of the National Academy of Sciences*, 114(26):E5070–E5076; DOI: 10.1073/pnas.1615978114.

Bechtel W. and Abrahamsen A. (2005). Explanation: A Mechanist Alternative. *Studies in the History and Philosophy of the Biological and Biomedical Sciences*, 36:421–441.

Bennett A. and Checkel J. T. (2014). Process Tracing. From Philosophical Roots to Best Practices. In Bennett A. and Checkel J. T. (eds.), *Process Tracing: From Metaphor to Analytic Tool*. Cambridge: Cambridge University Press.

Berk R. (2010). What You Can and Can't Properly Do with Regression. *Journal of Quantitative Criminology*, 26:481–487.

Berk R. A., Brown L., George E., Pitkin E., Traskin M., Zhang K., and Zhao L. (2013). What You Can Learn from Wrong Causal Models. In Morgan S. L. (ed.), *Handbook of Causal Analysis for Social Research*, pages 403–424. Dordrecht: Springer.

Bianchi F. and Squazzoni F. (2015). Agent-based Models in Sociology. *Computational Statistics*, 7(4):284–306.

Billari F. and Prskawetz A. (eds.) (2003). *Agent-Based Computational Demography: Using Simulation to Improve Our Understanding of Demographic Behaviour*. Heidelberg: Physica Verlag.

Billari F. C., Prskawetz A., Diaz B. A., and Fent T. (2007). The "Wedding-ring": An Agent-Based Marriage Model Based on Social Interaction. *Demographic Research*, 17(3):59–82.

Birks D., Townsley M., and Stewart A. (2012). Generative Explanations of Crime: Using Simulation to Test Criminological Theory. *Criminology*, 50(1):221–254.

Boero R., Bravo G., Castellani M., and Squazzoni F. (2010). Why Bother with What Others Tell You? An Experimental Data-Driven Agent-Based Model. *Journal of Artificial Societies and Social Simulation*, 13(3):6.

Boero R. and Squazzoni F. (2005). Does Empirical Embeddedness Matter? Methodological Issues on Agent-Based Models for Analytical Social Science. *Journal of Artificial Societies and Social Simulation*, 8(4):6.

Bollen K. A. (2012). Instrumental Variables in Sociology and the Social Sciences. *Annual Review of Sociology*, 38:37–72.

Bonabeau E. (2002). Agent-based Modeling: Methods and Techniques for Simulating Human Systems, *Proceedings of the National Academy of Sciences*, 99(suppl. 3):7280–7287.

Boudon R. (1974). *Education, Opportunity, and Social Inequality*. New York: Wiley.

Boudon R. (1976). Comment on Hauser's Review of Education, Opportunity, and Social Inequality. *American Journal of Sociology*, 81(5):1175–1187.

Boudon R. (1979). Generating Models as a Research Strategy. In Rossi P. H. (ed.), *Qualitative and Quantitative Social Research: Papers in Honor of Paul F. Lazarsfeld*, pages 51–64. New York: The Free Press.

Bound J., Jaeger D. A., and Baker R. M. (1995). Problems with Instrumental Variables Estimation When the Correlation between the Instruments and the Endogenous Explanatory Variable is Weak. *Journal of the American Statistical Association*, 90(430):443–450.

Bovens L. and Hartmann S. (2003). *Bayesian Epistemology*. Oxford: Oxford University Press.

Brady H. E. (2011). Causation and Explanation in Social Science. In Goodin R. E. (ed.), *The Oxford Handbook of Political Science*. Oxford: Oxford University Press.

Brashears M. E., Genkin M., and Suh C. S. (2017). In the Organization's Shadow: How Individual Behavior is Shaped by Organizational Leakage. *American Journal of Sociology*, 123(3):787–849.

Breen R. (2018). Some Methodological Problems in the Study of Multigenerational Mobility. *European Sociological Review*, 34(6):1 pages 603–611. https://doi.org/10.1093/esr/jcy037.

Breen R. (2022). Causal Inference and Estimation with Observational Data. (Ch. 14) In Gërxhani K., De Graaf N. D. and Raub W. (eds.), *Handbook of Sociological Science. Contributions to Rigorous Sociology*. Cheltenham (UK): Edward Elgar Publishing.

Breen R. and Karlson K. B. (2013). Counterfactual Causal Analysis and Nonlinear Probability Models. In Morgan S. L. (ed.), *Handbook of Causal Analysis for Social Research*, pages 167–187. Dordrecht: Springer.

Brenner T. and Werker C. (2007). A Taxonomy of Inference in Simulation Models. *Computational Economics*, 30:227–244.

Brown D. and Robinson D. (2006). Effects of Heterogeneity in Residential Preferences on an Agent-Based Model of Urban Sprawl. *Ecology and Society*, 11(1):46.

Bruch E. (2014). How Population Structure Shapes Neighborhood Segregation. *American Journal of Sociology*, 119(5):1221–1278.

Bruch E. and Atwell J. (2015). Agent-Based Models in Empirical Social Research. *Sociological Methods and Research*, 44(2):186–221.

Bruch E. and Mare R. (2006). Neighborhood Choice and Neighborhood Change. *American Journal of Sociology*, 112:667–709.

Bruch E. and Mare R. (2009). Preferences and Pathways to Segregation: Reply to Van De Rijt, Siegel, and Macy. *American Journal of Sociology*, 114:1181–1198.

Bryan M. L. and Jenkins S. P. (2015). Multilevel Modelling of Country Effects: A Cautionary Tale. *European Sociological Review*, 10.1093/esr/jcv059.

Buchanan A. L., Bessey S., William C., King M., Murray E. J., Friedman S., Halloran M. E., and Brandon D. L. Marshall. (2020). Disseminated Effects in Agent Based Models: A Potential Outcomes Framework and Application to Inform Pre-Exposure Prophylaxis Coverage Levels for HIV Prevention. *Americal Journal of Epidemiology*, 10.1093/aje/kwaa239.

Bundgaard H., Bundgaard J. S., Raaschou-Pedersen D. E. T. et al. (2021). Effectiveness of Adding A Mask Recommendation to Other Public Health Measures to Prevent SARS-CoV-2 Infection in Danish Mask Wearers: A Randomized Controlled Trial. *Annals of Internal Medicine*, 174:335–343.

Canali S. (2019). Evaluating Evidential Pluralism in Epidemiology: Mechanistic Evidence in Exposome Research. *History and Philosophy of the Life Sciences*, 41:4. https://doi.org/10.1007/s40656-019-0241-6.

Carley K. M. (2002). Computational Organization Science: A New Frontier. *PNAS*, 99(3):7257–7262.

Carrella E., Bailey R., and Madsen J. K. (2020). Calibrating Agent-Based Models with Linear Regressions. *Journal of Artificial Societies and Social Simulation*, 23(1):7. http://jasss.soc.surrey.ac.uk/23/1/7.html.

Cartwright N. (1999). *The Dappled World: A Study of the Boundaries of Science*. Cambridge: Cambridge University Press.

Cartwright N. (2004). Causation: One Word, Many Things. *Philosophy of Science*, 71:805–819.

Cartwright N. (2007a). Are RCTs the Gold Standard? Contingency and Dissent in Science, CPNSS, LSE. Technical Report 01/07.

Cartwright N. (2007b). *Hunting Causes and Using Them*. Cambridge: Cambridge University Press.

Cartwright N., Goldfinch A., and Howick J. (2010). Evidence-Based Policy: Where Is Our Theory of Evidence? *Journal of Children Services*, 4(4):6–14.

Cartwright N. and Hardie J. (2013). *Evidence-based Policy: A Practical Guide to Doing It Better*. Oxford: Oxford University Press.

Casini L. (2012). Causation: Many Words, One Thing? *Theoria*, 74:203–219.

Casini L. (2014). Not-so-minimal Models: Between Isolation and Imagination. *Philosophy of the Social Sciences*, 44(5):646–672.

Casini L., Illari P., Russo F., and Williamson J. (2011). Models for Prediction, Explanation and Control: Recursive Bayesian Networks. *Theoria*, 26(70):5–33.

Casini L. and Manzo G. (2016). Agent-based Models and Causality. A Methodological Appraisal. Linköping University Electronic Press. The IAS Working Paper Series, 2016:7.

Castellano C., Fortunato S., and Loreto V. (2009). Statistical Physics of Social Dynamics. *Reviews of Modern Physics*, 81:591–646.

Cederman L. E. (2005). Computational Models of Social Forms: Advancing General Process Theory. *American Journal of Sociology*, 110(4):864–893.

Chan T. W. and Bolliver V. (2013). The Grandparents Effect in Social Mobility: Evidence from British Birth Cohort Studies. *American Sociological Review*, 78:662–678.

Chattoe-Brown E. (2014). Using Agent Based Modelling to Integrate Data on Attitude Change. *Sociological Research Online*, 19(1):16.

Chattoe-Brown E., Gilbert N., Robertson D. A. and Watts C. (2021). *Reproduction as a Means of Evaluating Policy Models: A Case Study of a COVID-19 Simulation*, medRxiv 2021.01.29.21250743. doi: https://doi.org/10.1101/2021.01.29.21250743.

Chavali A. K., Gianchandani E. P., Tung K. S., Lawrence M. B., Peirce S. M., and Papin J. A. (2008). Characterizing Emergent Properties of Immunological Systems with Multi-cellular Rule-based Computational Modeling. *Trends in Immunology*, 29:589–599.

Cheng S. (2021). The Shifting Life Course Patterns of Wage Inequality, *Social Forces*, https://doi.org/10.1093/sf/soab003.

Christakis N. A. and Fowler J. H. (2013). Social Contagion Theory: Examining Dynamic Social Networks and Human Behavior. *Statistics in Medicine*, 32:556–577.

Clarke B., Leuridan B., and Williamson J. (2014). Modeling Mechanisms with Causal Cycles. *Synthese*, 191:1651–1681.

Claveau F. (2011). Evidential Variety as a Source of Credibility for Causal Inference: Beyond Sharp Designs and Structural Models. *Journal of Economic Methodology*, 18(3):233–253.

Claveau F. (2012). The Russo–Williamson Theses in the Social Sciences: Causal Inference Drawing on Two Types of Evidence. *Studies in History and Philosophy of Biological and Biomedical Sciences*, 43:806–813.

Cointet J.-P. and Roth C. (2007). How Realistic Should Knowledge Diffusion Models Be? *Journal of Artificial Societies and Social Simulation*, 10(3):5.

Colander D., Holt R. P. F., and Rosser J. B. (2004). *The Changing Face of Economics: Conversations with Cutting Edge Economists*. Ann Arbor, MI: University of Michigan Press.

Coleman J. S. (1986). Social Theory, Social Research, and a Theory of Action. *American Journal of Sociology*, 91:1309–1335.

Cox D. R. (1992). Causality: Some Statistical Aspects. *Journal of the Royal Statistical Society. Series A (Statistics in Society)*, 155(2):291–301.

Cox D. R. (2006). *Principles of Statistical Inference*. Cambridge: Cambridge University Press.

Cramer S. and Trimborn T. (2019). Stylized Facts and Agent-based Models. *ArXiv*, 1912.02684v1.

Craver C. and Bechtel W. (2007). Top-Down Causation Without Top-Down Causes. *Biology and Philosophy*, 22:547–563.

Craver C. F. (2007). *Explaining the Brain*. Oxford: Oxford University Press.

Davis P. and Lay-Yee R. (2019). *Simulating Social Change. Counterfactual Modeling for Social and Policy Inquiry*. New York: Springer International Publishing.

Dawid A. P. (2008). Beware of the DAG! *Journal of Machine Learning Research: Workshop and Conference Proceeding*, 6:59–86.

Dawid A. P., Faigman D. L., and Fienberg S. E. (2014). Fitting Science into Legal Contexts: Assessing Effects of Causes or Causes of Effects? *Sociological Methods & Research*, 43(3):359–390.

Daza S. and Kreuger L. K. (2019). Agent-Based Models for Assessing Complex Statistical Models: An Example Evaluating Selection and Social Influence Estimates from SIENA. *Sociological Methods & Research*, 1–38. 10.1177/0049124119826147.

De Grauwe P. (2010). Top-Down versus Bottom-Up Macroeconomics. *CESifo Economic Studies*, 56(4):465–497.

De Marchi S. and Page S. E. (2014). Agent-Based Models. *Annual Review of Political Science*, 17:1–20.

Deaton A. (2010). Instruments, Randomization, and Learning about Development. *Journal of Economic Literature*, 48:424–455.

Deaton A. and Cartwright N. (2018). Understanding and Misunderstanding Randomized Controlled Trials. *Social Science & Medecine*, 210:2–21.

Deffuant G., Weisbuch G., Amblard F., and Faure T. (2003). Simple Is Beautiful...and Necessary. *Journal of Artificial Societies and Social Simulation*, 6(1):6.

DellaPosta D., Shi Y., and Macy M. (2015). Why Do Liberals Drink Lattes. *American Journal of Sociology*, 120(5):1473–1511.

Delli Gatti D., Fagiolo G., Gallegati M., Richiardi M., and Russo A. (2018). *Agent-Based Models in Economics. A Toolkit.* Cambridge: Cambridge University Press.

Delre S. A., Jager W., Bijholt T. H. A., and Janssen M. A. (2010). Will It Spread or Not? the Effects of Social Influences and Network Topology on Innovation Diffusion. *Journal of Product Innovation Management*, 27:267–282.

Demeulenaere P. (ed.) (2011a). *Analytical Sociology and Social Mechanisms.* Cambridge: Cambridge University Press.

Demeulenaere P. (2011b). Introduction. In Demeulenaere P. (ed.), *Analytical Sociology and Social Mechanisms.* Cambridge: Cambridge University Press.

Demeulenaere P. (2011c). Causal Regularities, Action and Explanation. (Ch. 9) In Demeulenaere P. (ed.), *Analytical Sociology and Social Mechanisms.* Cambridge: Cambridge University Press, pages 173–199.

Di Iorio F. and Léon-Medina F. (2021). Analytical Sociology and Critical Realism. (Ch. 6) In Manzo G. (ed.), *Research Handbook on Analytical Sociology.* Cheltenham (UK): Edward Elgar.

Diez Roux A. V. (2015). The Virtual Epidemiologist—Promise and Peril. *American Journal of Epidemiology*, 181(2):100–102.

DiMaggio P. and Garip F. (2011). How Network Externalities Can Exacerbate Intergroup Inequality. *American Journal of Sociology*, 116(6):1887–1933.

Dollmann J. (2016). Less Choice, Less Inequality? A Natural Experiment on Social and Ethnic Differences in Educational Decision-Making. *European Sociological Review*, 31(2):203–215.

Doreian P. (1999). Causality in Social Network Analysis. *Sociological Methods and Research*, 30(1):81–114.

Drost C. J. and Vander Linden M. (2018). Toy Story: Homophily, Transmission and the Use of Simple Simulation Models for Assessing Variability in the Archaeological Record. *Journal of Archaeological Method and Theory*, 25(4):1087–1108.

Duffy J. (2006). Agent-Based Models and Human Subject Experiments. In Tesfatsion L. and Judd K. L. (eds.), *Handbook of Computational Economics. Agent-Based Computational Economics*, Vol. 2, pages 949–1011. Amsterdam: North-Holland Elsevier.

Dugundji E. R. and Gulyas L. (2008). Sociodynamic Discrete Choice on Networks in Space: Impacts of Agent Heterogeneity on Emergent Outcomes. *Environment and Planning B: Planning and Design*, 35:1028–1054.

Duo E., Glennerster R., and Kremer M. (2008). Using Randomization in Development Economics Research: A Toolkit. In Schultz T. P. and Strauss J. (eds.), *Handbook of Development Economic*, pages 3895–3962. Amsterdam and Oxford: Elsevier, North-Holland.

Durlauf S. N. and Ioannides Y. M. M. (2010). Social Interactions. *Annual Review of Economics*, 2:451–478.

Eberlen J., Scholz G., and Gagliolo M. (2017). Simulate This! An Introduction to Agent-Based Models and Their Power to Improve Your Research Practice. *International Review of Social Psychology*, 30(1):149–160.

Edmonds B. (2015). Using Qualitative Evidence to Inform the Specification of Agent-Based Models. *Journal of Artificial Societies and Social Simulation*, 18(1):18.

Edmonds B., Le Page C., Bithell M., Chattoe-Brown E., Grimm V., Meyer R., Montañola-Sales C., Ormerod P., Root H., and Squazzoni F. (2019). Different Modelling Purposes. *Journal of Artificial Societies and Social Simulation*, 22(3):6. http://jasss.soc.surrey.ac.uk/22/3/6.html.

Edmonds B. and Moss S. J. (2005). From KISS to KIDS—An 'Antisimplistic' Modelling Approach. In Davidson P. et al. (ed.), *Multi Agent Based Simulation 2004, Volume 3415 of Lecture Notes in Artificial Intelligence*, pages 130–144. Berlin: Springer.

Elsenbroich C. (2012). Explanation in Agent-Based Modelling: Functions, Causality or Mechanisms? *Journal of Artificial Societies and Social Simulation*, 15(3):1. http://jasss.soc.surrey.ac.uk/15/3/1.html.

Elster J. (2009a). Excessive Ambitions. *Capitalism and Society*, 4:2. 10.2202/1932-0213.1055.

Elster J. (2009b). *Alexis De Tocqueville: The First Social Scientist*. New York: Cambridge University Press.

Elwert F. (2013). Graphical Causal Models. In Morgan S. L. (ed.), *Handbook of Causal Analysis for Social Research*, pages 245–273. Dordrecht: Springer.

Elwert F. and Winship C. (2015). Endogenous Selection Bias: The Problem of Conditioning on a Collider Variable. *Annual Review of Sociology*, 40:31–53.

Epstein J. M. (1999). Agent-Based Computational Models and Generative Social Science. *Complexity*, 4(5):41–60.

Epstein J. M. (2006). *Generative Social Science: Studies in Agent-Based Computational Modeling*. Princeton: Princeton University Press.

Epstein J. M. and Axtell R. (1996). *Growing Artificial Societies. Social Science from the Bottom Up*. Cambridge, MA: MIT Press.

Ermakoff I. (2019). Causality and History: Modes of Causal Investigation in Historical Social Sciences. *Annual Review of Sociology*, 45(1):581–606.

Fagiolo G., Moneta A., and Windrum P. (2007). A Critical Guide to Empirical Validation of Agent-Based Models in Economics: Methodologies, Procedures, and Open Problems. *Computational Economics*, 30:195–226.

Fararo T. J. (1969a). The Nature of Mathematical Sociology. *Social Research*, 36:75–92.

Fararo T. J. (1969b). Stochastic Processes. In Borgatta E. F. (ed.), *Sociological Methodology*. San Francisco: Jossey-Bass.

Fararo T. J. and Butts C. T. (1999). Advance in Generative Structuralism: Structured Agency and Multilevel Dynamics. *Journal of Mathematical Sociology*, 24:1–65.

Fararo T. J. and Kosaka K. (1976). A Mathematical Analysis of Boudon's IEO Model. *Social Science Information*, 15(2-3):431–475.

Farmer J. D. and Foley D. (2009). The Economy Needs Agent-Based Modelling. *Nature*, 460:685–686.

Ferber J., Michel F., and Baez J. (2005). AGRE: Integrating Environments with Organizations. In Weyns D., Parunak V. D. and Michel F. (eds.), *Environments for Multi-Agent Systems*, pages 44–56. Berlin: Springer.

Fioretti G. (2013). Agent-based Simulation Models in Organization Science. *Organizational Research Methods*, 16:227–242.

Flache A. and De Matos Fernandes C. A. (2021). Agent-based Computational Models. (Ch. 24) In Manzo G. (ed.), *Research Handbook on Analytical Sociology*. Cheltenham (UK): Edward Elgar.

Flache A., Mäs M., Feliciani T., Chattoe-Brown E., Deffuant G., Huet S., and Lorenz J. (2017). Models of Social Influence: Towards the Next Frontiers. *Journal of Artificial Societies and Social Simulation*, 20(4):2.

Flache A., Mäs M., and Keijzer M. A. (2022). Computational Approaches in Rigorous Sociology: Agent-based Computational Sociology and Computational Social Science. (Ch. 4) In Gërxhani K., De Graaf N. D. and Raub W. (eds.), *Handbook of Sociological Science. Contributions to Rigorous Sociology*. Cheltenham (UK): Edward Elgar.

Fountain C. and Stovel K. (2014). Turbulent Careers: Social Networks, Employer Hiring Preferences, and Job Instability. In Manzo G. (ed.), *Analytical Sociology: Actions and Networks*, pages 339–370. Chichester (UK): Wiley.

Freedman D. A. (1995). Some Issues in the Foundations of Statistics: Probability and Statistical Models. *Foundations of Science*, 1:19–39.

Freedman D. A. (2005). Linear Statistical Models for Causation: A Critical Review. In Everitt B. and Howell D. C. (eds.), *Encyclopedia of Statistics in Behavioral Science*. Hoboken, NJ: Wiley

Freedman D. A. (2009). *Statistical Models: Theory and Practice*. Cambridge: Cambridge University Press.

Freedman D. A. (2010). *Statistical Models and Causal Inference: A Dialogue with the Social Sciences*. Cambridge: Cambridge University Press.

Freedman D. A. and Humphreys P. (1996). The Grand Leap. *British Journal for the Philosophy of Science*, 47:113–123.

Freedman D. A. and Humphreys P. (1999). Are There Algorithms That Discover Causal Structure? *Synthese*, 121:29–54.

Freese J. and Kevern J. A. (2013). Types of Causes. In Morgan S. L. (ed.), *Handbook of Causal Analysis for Social Research*, pages 27–41. Dordrecht: Springer

Frey V. and van de Rijt, A. (2016). Arbitrary Inequality in Reputation Systems. *Scientific Reports*, 6:38304, https://doi.org/10.1038/srep38304.

Frias-Martinez E., Williamson G., and Frias-Martinez V. (2011). An Agent-Based Model of Epidemic Spread Using Human Mobility and Social Network Information. In *Privacy, Security, Risk and Trust (PASSAT)* and *2011 IEEE Third International Conference on Social Computing (Social Com)*, pages 57–64.

Frigg R. and Reiss J. (2009). The Philosophy of Simulation: Hot New Issues or Same Old Stew? *Synthese*, 169(3):593–613.

Gabbriellini S. and Torroni P. (2014). Arguments in Social Networks. In *Proceedings of the 2013 International Conference on Autonomous Agents and Multi-agent Systems (AAMAS'13)*, pages 119–1120.

Gallegati M. and Kirman A. P. (eds) (1999). *Beyond the Representative Agent*. Aldershot and Lyme, NH: Edward Elgar.

Gangl M. (2010). Causal Inference in Sociological Research. *Annual Review of Sociology*, 36:21–47.

Gangl M. (2013). Partial Identification and Sensitivity Analysis. In Morgan S. L. (ed.), *Handbook of Causal Analysis for Social Research*, pages 377–402. Dordrecht: Springer.

Gelman A. (2011). Causality and Statistical Learning. *American Journal of Sociology*, 117(3):955–966.

Gelman A. and Basbøll T. (2014). When Do Stories Work? Evidence and Illustration in the Social Sciences. *Sociological Methods & Research*, 43(4):547–570.

Gelman A. and Hill J. (2007). *Data Analysis Using Regression and Multilevel/Hierarchical Models*. Cambridge: Cambridge University Press.

Gelman A. and Imbens G. (2013). Why Ask Why? Forward Causal Inference and Reverse Causal Questions. National Bureau of Economic Research Working Paper Series, No. 19614 (10.3386/w19614).

Gelman A. and Loken E. (2014). The Statistical Crisis in Science. *American Scientist,* 102(November–December):460–465.

Gelman A. and O'Rourke K. (2013). Convincing Evidence. *arXiv,* 1307.

Gerring J. (2008). The Mechanismic Worldview: Thinking Inside the Box. *British Journal of Political Science,* 38(1):161–179.

Ghorbani A., Dijkema G., and Schrauwen N. (2015). Structuring Qualitative Data for Agent-Based Modelling. *Journal of Artificial Societies and Social Simulation,* 18(1):2.

Gilbert N. and Abbott A. (eds.) (2005). Social Science Computation. *American Journal of Sociology,* 110(4).

Gilbert N. and Troitzsch K. (2005). *Simulation for the Social Scientist.* Maidenhead (UK): Open University Press.

Gintis H. (2009). *The Bounds of Reason: Game Theory and the Unification of the Behavioral Sciences.* Princeton: Princeton University Press.

Gintis H. (2013). Markov Models of Social Dynamics: Theory and Applications. *ACM Transactions on Intelligent Systems and Technology,* 4(3):53.

Glennan S. (1996). Mechanisms and the Nature of Causation. *Erkenntnis,* 44:49–71.

Glennan S. (2002). Rethinking Mechanistic Explanation. *Philosophy of Science,* 69(3):S342–S353.

Glymour C., Zhang K., and Spirtes P. (2019). Review of Causal Discovery Methods Based on Graphical Models. *Frontiers in Genetics,* 10:1–15.

Glymour M. and Greenland S. (2008). Causal Diagrams. In Rothman K., Greenland S. and Lash T. (eds.), *Modern Epidemiology,* pages 183–209. Philadelphia: Lippincott Williams & Wilkins.

Goldthorpe J. H. (2001). Causation, Statistics and Sociology. *European Sociological Review,* 17(1):1–20.

Goldthorpe J. H. (2016). *Sociology as a Population Science.* Cambridge: Cambridge University Press.

Goldthorpe J. H. (2021). *Pioneers of Sociological Science: Statistical Foundations and the Theory of Action.* Cambridge: Cambridge University Press.

Gonzales-Bailon S. and Murphy T. E. (2013). Social Interactions and Long-Term Fertility Dynamics. A Simulation Experiment in the Context of the French Fertility Decline. *Population Studies,* 67(2):135–155.

Gorski P. (2009). Social Mechanisms and Comparative-Historical Sociology: A Critical Realist Proposal. In Hedström P. and Wittrock B. (eds.), *Frontiers of Sociology,* pages 147–194. Leiden (Netherlands): Brill.

Gould R. V. (2002). The Origins of Status Hierarchies: A Formal Theory and Empirical Test. *American Journal of Sociology,* 107(5):1143–1178.

Granger C. W. J. (1969). Investigating Causal Relations by Econometric Models and Cross-Spectral Methods. *Econometrica,* 37(3):424–438.

Greenland S. (2010). Overthrowing the Tyranny of Null Hypotheses Hidden in Causal Diagrams. In Dechter R., Geffner H., and Halpern J. Y. (eds.), *Heuristics, Probability and Causality: A Tribute to Judea Pearl,* pages 365–382. London: College Publications.

Grimm V., Berger U., Bastiansen F., Eliassen S., Ginot V., Giske J., Goss-Custard J., Grand T., Heinz S. K., Huse G., Huth A., Jepsen J. U., Jørgensen C., Mooij W. M., Mër G., Piou C., Railsback S. F., Robbins A. M., Robbins Muller B., Pe´M M., Rossmanith E., Ruger N., Strand E., Souissi S., Stillman R. A., Vabø R., Visser U., and DeAngelis D. L. (2006). A Standard Protocol for Describing Individual-Based and Agent-Based Models. *Ecological Modelling*, 198(1–2):115–126.

Gross N. (2009). A Pragmatist Theory of Social Mechanisms. *American Sociological Review*, 74:358–379.

Gross N. (2018). The Structure of Causal Chains. *Sociological Theory*, 36(4):343–367.

Grossman G. and Baldassarri D. (2012). The Impact of Elections on Cooperation: Evidence from a Lab in the Field Experiment in Uganda. *American Journal of Political Science*, 56(4):964–985.

Grüne-Yanoff T. (2009a). The Explanatory Potential of Artificial Societies. *Synthese*, 169(3):539–555.

Grüne-Yanoff T. (2009b). Learning from Minimal Economic Models. *Erkenntnis*, 70(1):81–99.

Guala F. (2002). Models, Simulations, and Experiments. In Magnani L. and Nersessian N. (eds.), *Model-Based Reasoning: Science, Technology, Values*, pages 59–74. New York: Kluwer.

Hägerstrand T. (1965). A Montecarlo Approach to Diffusion. *European Journal of Sociology*, 6(1):43–67.

Hall N. (2004). Two Concepts of Causation. In Collins J., Hall N. and Paul L. A. (eds.), *Causation and Counterfactuals*, pages 225–276. Cambridge, MA: MIT Press.

Halloran M. E., Auranen K., Baird S. et al. (2017). Simulations for Designing and Interpreting Intervention Trials in Infectious Diseases. *BMC Medicine*, 15:223. https://doi.org/10.1186/s12916-017-0985-3.

Halloran M. E. and Hudgens M. G. (2016). Dependent Happenings: A Recent Methodological Review. *Current Epidemiology Reports*, 3(4):297–305. 10.1007/s40471-016-0086-4.

Hansen L. P. and Heckman J. J. (1996). The Empirical Foundations of Calibration. *Journal of Economic Perspectives*, 10(1):87–104.

Harré R. (1972). *The Philosophies of Sciences. An Introduction Survey*. Oxford: Oxford University Press.

Harrington J. J. and Chang M.-H. (2005). Co-evolution of Firms and Consumers and the Implications for Market Dominance. *Journal of Economic Dynamics and Control*, 29(1–2):245–276.

Hassan S., Pavon J., Antunes L., and Gilbert N. (2010). Injecting Data into Agent-Based Simulation. In Takadama K., Deffuant G. and Cioffi-Revilla C. (eds.), *Simulating Interacting Agents and Social Phenomena: The Second World Congress Springer, Tokyo (2010), Volume 7 of Springer Series on Agent Based Social Systems*, pages 179–191. Tokyo: Springer.

Hauser R. (1976). Review Essay. On Boudon's Model of Social Mobility. *American Journal of Sociology*, 8(1):911–928.

Hausman D. (1992). *The Inexact and Separate Science of Economics*. Cambridge: Cambridge University Press.

Hayward S. (2006). Agent-based Modelling with Wavelets and an Evolutionary Artificial Neural Network: Applications to CAC 40 Forecasting. In Chatterjee E., Chakrabarti E. and Bikas K. R. (eds.), *Econophysics of Stock and Other Markets*, pages 163–174. Milan: Springer.

Heckbert S., Baynes T., and Reeson A. (2010). Agent-Based Modeling in Ecological Economics. *Annals of the New York Academy of Sciences*, 1185:39–53.

Heckman J. J. (2005). The Scientific Model of Causality. *Sociological Methodology*, 35(1):1–97.

Hedstrom P. (2009). Studying Mechanisms to Strengthen Causal Inferences in Quantitative Research. In Box-Steffensmeier J. M., Brady H. E., and Collier D. (eds.), *The Oxford Handbook of Political Methodology*, pages 319–335. Oxford: Oxford University Press.

Hedström P. (2005). *Dissecting the Social. On the Principles of Analytical Sociology*. Cambridge: Cambridge University Press.

Hedström P. (2021). The Past and the Future of Analytical Sociology. In Manzo G. (ed.), *Research Handbook on Analytical Sociology*. Cheltenham (UK): Edward Elgar.

Hedström P. and Bearman P. (eds.) (2009). *The Oxford Handbook of Analytical Sociology*. Oxford: Oxford University Press.

Hedström P. and Manzo G. (eds.) (2015). Agent-Based Modeling: Advances and Challenges. *Sociological Methods and Research*, 44(2).

Hedström P. and Swedberg R. (1998). Social Mechanisms: An Introductory Essay. In Hedström P. and Swedberg R. (eds.), *Social Mechanisms: An Analytical Approach to Social Theory*, pages 1–31. Cambridge: Cambridge University Press.

Hedström P. and Ylikoski P. (2010). Causal Mechanisms in the Social Sciences. *Annual Review of Sociology*, 36:49–67.

Hegselmann R. (2017). Thomas C. Schelling and James M. Sakoda: The Intellectual, Technical, and Social History of a Model. *Journal of Artificial Societies and Social Simulation*, 20(3):http://jasss.soc.surrey.ac.uk/20/3/15.html.

Helbing D. (2012). *Social Self-Organization. Agent-based Simulations and Experiments to Study Emergent Social Behavior*. Berlin: Springer.

Hernán M. A. and Robins J. M. (2020). *Causal Inference: What If*. Boca Raton, FL: Chapman & Hall/CRC (https://www.hsph.harvard.edu/miguel-hernan/causal-inference-book).

Hesslow G. (1976). Discussion: Two Notes on the Probabilistic Approach to Causality. *Philosophy of SCIENCE*, 43:290–292.

Hjorth A., Head B., Brady C., and Uri W. (2020). LevelSpace: A NetLogo Extension for Multi-Level Agent-Based Modeling. *Journal of Artificial Societies and Social Simulation*, 23(1):4. http://jasss.soc.surrey.ac.uk/23/1/4.html.

Holland P. W. (1986). Statistics and Causal Inference. *Journal of the American Statistical Association*, 81(396):945–960.

Holm S., Hilty L. M., Lemm R., and Thees O. (2018). Empirical Validation of an Agent-based Model of Wood Markets in Switzerland. *PLoS One*, 13(1): e0190605.https://doi.org/10.1371/journal.pone.0190605.

Hong G. and Raudenbush S. W. (2013). Heterogeneous Agents, Social Interactions, and Causal Inference. In Morgan S. L. (ed.), *Handbook of Causal Analysis for Social Research*, pages 331–352. Dordrecht: Springer.

Hoover K. (2008a). Does Macroeconomics Need Microfoundations? In Hausman D. (ed.), *Philosophy of Economics*, pages 315–333. Cambridge: Cambridge University Press.

Hoover K. (2008b). Microfoundations and the Ontology of Macroeconomics. In Ross D. and Kincaid H. (eds.), *Oxford Handbook of Philosophy of Economics*, pages 386–409. Oxford: Oxford University Press.

Hoover K. (2012). Causal Structure and Hierarchies of Models. *Studies in History and Philosophy of Biological and Biomedical Sciences*, 43:778–786.

Hummon N. P. and Fararo T. J. (1995a). Actors and Networks as Objects. *Social Networks*, 17:1–26.

Hummon N. P. and Fararo T. J. (1995b). The Emergence of Computational Sociology. *Journal of Mathematical Sociology*, 20(2-3):79–87.

Humphreys P. (2009). The Philosophical Novelty of Computer Simulation Methods. *Synthese*, 169(3):615–626.

Illari P. (2011). Mechanistic Evidence: Disambiguating the Russo–Williamson Thesis. *International Studies in the Philosophy of Science*, 25(2):1–19.

Imbens G. W. and Rubin D. B. (2015). *Causal Inference for Statistics, Social, and Biomedical Sciences*. Cambridge: Cambridge University Press.

Izquierdo L. R., Izquierdo S. S., Galàn J. M., and Santos J. I. (2009). Techniques to Understand Computer Simulations: Markov Chain Analysis. *Journal of Artificial Societies and Social Simulation*, 12(1):6.

Janssen M. A. and Jager W. (2001). Fashions, Habits and Changing Preferences: Simulation of Psychological Factors Affecting Market Dynamics. *Journal of Economic Psychology*, 22:745–772.

Janssen M. A. and Jager W. (2003). Simulating Market Dynamics: Interactions Between Consumer Psychology and Social Networks. *Artificial Life*, 9:343–356.

Kaidesoja T. (2021a). Causal Inference and Modeling. In McIntyre L. and Rosenberg A. (eds.). *Routledge Companion to Philosophy of Social Science*. London: Routledge, Forthcoming.

Kaidesoja T. (2021b). Three Concepts of Causal Mechanism in the Social Sciences. In Erola J. and Naumanen P. (eds.), *Norms, Moral and Social Structures*. Forthcoming.

Kalisch M., Mächler M., Colombo D., Maathuis M. H., and Bühlmann P. (2012). Causal Inference Using Graphical Models with the R Package Pcalg. *Journal of Statistical Software*, 47:11.

Kalter F. and Kroneberg C. (2014). Between Mechanism Talk and Mechanism Cult: New Emphases in Explanatory Sociology and Empirical Research. *Kolner Zeitschrift Fur Soziologie Und Sozialpsychologie*, 66:S91–S115.

Kant J.-D., Ballot G., and Goudet O. (2020). WorkSim: An Agent-Based Model of Labor Markets. *Journal of Artificial Societies and Social Simulation*, 23(4):4. http://jasss.soc.surrey.ac.uk/23/4/4.html.

Karlsson G. (1958). *Social Mechanisms: Studies in Sociological Theory*. Stockholm: Free Press.

Karpiński Z. and Skvoretz J. (2015). Repulsed by the 'Other': Integrating Theory with Method in the Study of Intergroup Association. *Sociological Theory*, 33:20–43.

Keuchenius A., Törnberg P., and Uitermark J. (2021). Adoption and Adaptation: A Computational Case Study of the Spread of Granovetter's Weak Ties Hypothesis. *Social Networks*, 66:10–25.

Kirman A. (1992). Whom or What Does the Representative Individual Represent? *Journal of Economic Perspectives*, 6(2):117–136.

Kistler M. (2002). Causation in Contemporary Analytical Philosophy. In Esposito C. and Porro P. (eds.), *Quaestio. Annuario Di Storia Della Metafisica*, Vol. 2, pages 635–668. Turnhout Belgium: Brepols.

Knight C. R. and Winship C. (2013). The Causal Implications of Mechanistic Thinking: Identification Using Directed Acyclic Graphs (DAGs). In Morgan S. L. (ed.), *Handbook of Causal Analysis for Social Research*, pages 275–299. Dordrecht: Springer.

Korb K. B. and Nicholson A. E. (2011). *Bayesian Artificial Intelligence*. Boca Raton, FL: CRC Press, 2nd edition.

Kreager D. A., Young J. T. N., Haynie D. L., Bouchard M., Schaefer D. R., and Zajac G. (2017). Where "Old Heads" Prevail: Inmate Hierarchy in a Men's Prison Unit. *American Sociological Review*, 82(4):685–718.

Kruse H. (2017). The SES-Specific Neighbourhood Effect on Interethnic Friendship Formation. The Case of Adolescent Immigrants in Germany. *European Sociological Review*, 33(2):182–194.

Landes J. (2020). The Variety of Evidence Thesis and Its Independence of Degrees of Independence. *Synthese*, https://doi.org/10.1007/s11229-020-02738-5.

Law A. M. (2007). *Simulation Modeling and Analysis*. New York: McGraw-Hill.

Leombruni R. and Richiardi M. (2005). Why are Economists Skeptical about Agent-Based Simulations? *Physica A*, 355:103–109.

León-Medina F. J. (2017). Analytical Sociology and Agent-Based Modeling: Is Generative Sufficiency Sufficient? *Sociological Theory*, 35(3):157–178.

Lipton P. (2004). *Inference to the Best Explanation*. London: Routledge, 2nd edition.

Lipton P. (2009). Causation and Explanation. In Beebee H., Menzies P. and Hitchcock C. (eds.), *The Oxford Handbook of Causation*, pages 619–631. Oxford: Oxford University Press.

Little D. (2012). Analytical Sociology and the Rest of Sociology. *Sociologica*, 1:1–47.

Little D. (2016). *New Directions in the Philosophy of Social Science*. London: Rowman & Littlefield.

Longworth F. (2006). Causation, Pluralism and Moral Responsibility. *Philosophica*, 77(1):45–68.

Lucas R. (1976). Econometric Policy Evaluation: A Critique. In Brunner K. and Meltzer A. (eds.), *The Phillips Curve and Labor Markets*, volume 1 of Carnegie-Rochester Conference. Series on Public Policy, pages 19–46. Amsterdam: North-Holland.

Lux T. and Marchesi M. (1999). Scaling and Criticality in a Stochastic Multi-Agent Model of a Financial Market. *Nature*, 397:498–500.

Machamer P., Darden L., and Craver C. (2000). Thinking about Mechanisms. *Philosophy of Science*, 67:1–25.

Macy M. and Flache A. (2009). Social Dynamics from the Bottom-Up: Agent-based Models of Social Interaction. In Hedström P. and Bearman P. (eds.), *The Oxford Handbook of Analytical Sociology*, CCCCh Ch. 11. Oxford: Oxford University Press.

Macy M. and Sato Y. (2008). Reply to Will and Hegselmann. *Journal of Artificial Societies and Social Simulation*, 11(4):11.

Macy M. W. and Willer R. (2002). From Factors to Actors: Computational Sociology and Agent-Based Modeling. *Annual Review of Sociology*, 28:143–166.

Magliocca N. R., Brown D. G., and Ellis E. C. (2014). Cross-Site Comparison of Land-Use Decision-Making and Its Consequences across Land Systems with a Generalized Agent-Based Model. *PLoS One*, 9(1):e86179.

Mahoney J. (2000). Strategies of Causal Inference in Small-N Analysis. *Sociological Methods and Research*, 28(4):387–424.

Mahoney J. (2001). Beyond Correlational Analysis: Recent Innovations in Theory and Method. *Sociological Forum*, 16(3):575–593.

Mahoney J. (2008). Toward a Unified Theory of Causality. *Comparative Political Studies*, 41(4–5):412–436.

Mahoney J., Goertz G., and Ragin C. C. (2013). Causal Models and Counterfactuals. In Morgan S. L. (ed.), *Handbook of Causal Analysis for Social Research*, pages 75–90. Dordrecht: Springer.

Makovi K. and Winship C. (2021). Advances in Mediation Analysis. In Manzo G. (ed.), *Research Handbook on Analytical Sociology*, Ch. 20. Cheltenham (UK): Edward Elgar

Manski C. F. (2003). *Partial Identification of Probability Distributions*. New York: Springer.

Manski C. F. (2007). *Identification for Prediction and Decision*. Cambridge, MA: Harvard University Press.

Manski C. F. (2013). Identification of Treatment Response with Social Interactions. *Econometrics Journal*, 16(1):S1–S23.

Manzo G. (2007). Variables, Mechanisms, and Simulations: Can the Three Methods Be Synthesized? *Revue Française De Sociologie*, 48(5):35–71.

Manzo G. (2010). Analytical Sociology and Its Critics. *European Journal of Sociology*, 51(1):129–170.

Manzo G. (2013). Educational Choices and Social Interactions: A Formal Model and A Computational Test. *Comparative Social Research*, 30:47–100.

Manzo G. (2014a). Data, Generative Models, and Mechanisms: More on the Principles of Analytical Sociology. In Manzo G. (ed.), *Analytical Sociology: Actions and Networks*, pages 4–52. Chichester (UK): Wiley.

Manzo G. (2014b). The Potential and Limitations of Agent-based Simulation: An Introduction. *Revue Française De Sociologie*, 55(4):653–688. 10.3917/rfs.554.0653.

Manzo G. (2020). Agent-based Models and Methodological Individualism: Are They Fundamentally Linked? *L'année Sociologique*, 70(1):197–229.

Manzo G. (2021). Does Analytical Sociology Practice What It Preaches? An Assessment of Analytical Sociology through the Merton Award. (Introduction) In Manzo G. (ed.), *Research Handbook on Analytical Sociology*. Cheltenham (UK): Edward Elgar.

Manzo G. and Baldassarri D. (2015). Heuristics, Interactions, and Status Hierarchies: An Agent-based Model of Deference Exchange. *Sociological Methods and Research*, 44(3):329–387.

Manzo G., Gabbriellini S., Roux V., and M'Mbogori F. N. (2018). Complex Contagions and the Diffusion of Innovations: Evidence from a Small-N Study. *Journal of Archaeological Method and Theory*, 25(4):1109–1154.

Manzo G. and van de Rijt A. (2020). Halting SARS-CoV-2 by Targeting High-Contact Individuals. *Journal of Artificial Societies and Social Simulation*, 23(4):10. http://jasss.soc.surrey.ac.uk/23/4/10.html.

Marini M. and Singer B. (1988). Causality in the Social Sciences. In *Sociological Methodology*, pages 347–409. San Francisco: Jossey-Bass.

Murray E. J., Robins J. M., Seage G. R. III, Freedberg K. A. and Hernan M. A. (2017). A Comparison of Agent-based Models and the Parametric G-Formula for Causal Inference. *American Journal of Epidemiology*, 186(2):131–142.

Marshall B. D. and Galea S. (2015). Formalizing the Role of Agent-Based Modeling in Causal Inference and Epidemiology. *American Journal of Epidemiology*, 181(2):92–99.

Mäs M. and Flache A. (2013). Differentiation without Distancing. Explaining BiPolarization of Opinions without Negative Influence. *PLoS One*, 8(11):e74516.

Mathieu P., Beaufils B., and Brandouy O. (2005). *Artificial Economics: Agent-based Methods in Finance, Game Theory and Their Applications*, Vol. 564. Berlin: Springer Science & Business Media.

Mayo D. G. (2018). *Statistical Inference as Severe Testing. How to Get beyond the Statistical Wars*. Cambridge: Cambridge University Press.

Melamed D. and Savage S. V. (2016). Status, Faction Sizes, and Social Influence: Testing the Theoretical Mechanism. *American Journal of Sociology*, 122(1):201–232.

Menzies P. (2012). The Causal Structure of Mechanisms. *Studies in History and Philosophy of Biological and Biomedical Sciences*, 43:796–805.

Mill J. S. (1882). *A System of Logic, Ratiocinative and Inductive*. New York: Harper & Brothers, 8th edition [Project Gutenberg EBook, Ebook 27942].

Miller J. H. and Page S. E. (2004). The Standing Ovation Problem. *Complexity*, 9(5):8–16.

Miller J. H. and Page S. E. (2007). *Complex Adaptive Systems: An Introduction to Computational Models of Social Life*. Princeton: Princeton University Press.

Mitze T., Kosfeld R., Rode J., and Wälde K. (2020). Face Masks Considerably Reduce COVID-19 Cases in Germany. *Proceedings of the National Academy of Sciences*, 117(51):32293–32301.

Moneta A. and Russo F. (2014). Causal Models and Evidential Pluralism in Econometrics. *Journal of Economic Methodology*, 21(1):54–76.

Morgan M. S. (2003). Experiments without Material Intervention: Model Experiments, Virtual Experiments and Virtually Experiments. In Radder H. (ed.), *The Philosophy of Scientific Experimentation*, pages 217–235. Pittsburgh, PA: University of Pittsburgh Press.

Morgan M. S. (2012). *The World in the Model: How Economists Work and Think*. Cambridge: Cambridge University Press.

Morgan S. L. (2005). *On the Edge of Commitment: Educational Attainment and Race in the United States*. Stanford, CA: Stanford University Press.

Morgan S. L. (ed.) (2013). *Handbook of Causal Analysis for Social Research*. Berlin: Springer.

Morgan S. L. and Winship C. (2015). *Counterfactuals and Causal Inference: Methods and Principles for Social Research*. Cambridge: Cambridge University Press, 2nd edition.

Morrison M. (2015). *Reconstructing Reality: Models, Mathematics, and Simulations*. Oxford: Oxford University Press.

Morvan G. (2013). Multi-level Agent-based Modeling—A Literature Survey. *ArXiv*, 1205.0561v7.

Mouchart M. and Russo F. (2011). Causal Explanation: Recursive Decompositions and Mechanisms. In Illari P., Russo F. and Williamson J. (eds.), *Causality in the Sciences*, pages 317–337. Oxford: Oxford University Press.

Muldoon R. (2007). Robust Simulations. *Philosophy of Science*, 74(5):873–883.

Muthukrishna M. and Schaller M. (2020). Are Collectivistic Cultures More Prone to Rapid Transformation? Computational Models of Cross-Cultural Differences, Social Network Structure, Dynamic Social Influence, and Cultural Change. *Personality and Social Psychology Review*, 24(2):103–120.

Nikolai C. and Madey G. (2009). Tools of the Trade: A Survey of Various Agent Based Modeling Platforms. *Journal of Artificial Societies and Social Simulation*, 12(2):2.

O'Sullivan D. (2008). Geographical Information Science: Agent-Based Models. *Progress in Human Geography*, 32:541–550.

O'Sullivan D. and Perry G. L. W. (2013). *Spatial Simulation: Exploring Pattern and Process*. Chichester (UK): Wiley.

Olsson E. (2017). Coherentist Theories of Epistemic Justification. In Zalta E. N. (ed.), *Stanford Encyclopedia of Philosophy*. In Stanford: Center for the Study of Language and Information (The Metaphysics Research Lab), Stanford University.

Opp K. (2007). Peter Hedström: Dissecting the Social. On the Principles of Analytical Sociology. *European Sociological Review*, 23:115–122.

Opp K. (2013). What Is Analytical Sociology? Strengths and Weaknesses of a New Sociological Research Program. *Social Science Information*, 52(3):329–360.

Oreskes N., Shrader-Frechette K., and Belitz K. (1994). Verification, Validation, and Confirmation of Numerical Models in the Earth Sciences. *Science*, 263(5147):641–646.

Page S. (2018). *The Model Thinker. What You Need to Know to Make Data Work for You*. New York: Basic Books.

Page S. E. (2008). Agent Based Models. In Blume L. E. and Durlauf S. N. (eds.), *The New Palgrave Dictionary of Economics*. Basingstoke (UK): Palgrave Macmillan, 2nd edition.

Parker J. and Epstein J. (2011). A Distributed Platform for Global-Scale Agent-Based Models of Disease Transmission. *ACM Transactions Modeling and Computer Simulations*, 22(1):2–33.

Parker W. (2008). Computer Simulation through an Error-Statistical Lens. *Synthese*, 163:371–384.

Parker W. (2009). Does Matter Really Matter? Computer Simulations, Experiments, and Materiality. *Synthese*, 169:483–496.

Pawson R. (1989). *A Measure for Measures: A Manifesto for Empirical Sociology*. London: Routledge.

Pearl J. (1993). Comment: Graphical Models, Causality, and Interventions. *Statistical Science*, 8(3):266–269.

Pearl J. (1995). Causal Diagrams for Empirical Research. *Biometrika*, 82(4):669–710.

Pearl J. (2009). *Causality: Models, Reasoning, and Inference*. Cambridge: Cambridge University Press.

Pearlman J. (2018). Gender Differences in the Impact of Job Mobility on Earnings: The Role of Occupational Segregation. *Social Science Research*, 74:30–44. https://doi.org/10.1016/j.ssresearch.2018.05.010.

Pinyol I. and Sabater-Mir J. (2013). Computational Trust and Reputation Models for Open Multi-agent Systems: A Review. *Artificial Intelligence Review*, 40(1):1–25.

Psillos S. (2007). What Is Causation? In Choksi B. In Natarajan C. (ed.), *The Episteme Reviews: Research Trends in Science, Technology and Mathematics Education*, pages 11–34. Bangalore: Macmillan India.

Quintana R. (2020). The Structure of Academic Achievement: Searching for Proximal Mechanisms Using Causal Discovery Algorithms. *Sociological Methods & Research*, 1–50. 10.1177/0049124120926208.

Railsback S. F. and Grimm V. (2019). *Agent-Based and Individual-Based Modeling: A Practical Introduction*. Princeton: Princeton University Press, 2nd edition.

Raub W., Buskens V., and Van Assen M. A. L. M. (2011). Micro-Macro Links and Microfoundations in Sociology. *Journal of Mathematical Sociology*, 35(1–3):1–25.

Reiss J. (2009). Causation in the Social Sciences. Evidence, Inference, and Purpose. *Philosophy of the Social Sciences*, 39(1):20–40.

Reiss J. (2011a). A Plea for (Good) Simulations: Nudging Economics toward an Experimental Science. *Simulation & Gaming*, 42(2):243–264.

Reiss J. (2011b). Third Time's a Charm: Causation, Science and Wittgensteinian Pluralism. In Illari P., Russo F. and Williamson J. (eds.), *Causality in the Sciences*, pages 907–927. Oxford: Oxford University Press.

Reiss J. (2013). *Philosophy of Economics. A Contemporary Introduction*. New York: Routledge.

Reiss J. (2015). A Pragmatist Theory of Evidence. *Philosophy of Science*, 82:341–362.

Rolfe M. (2014). Social Networks and Agent-Based Models. In Manzo G. (ed.), *Analytical Sociology: Actions and Networks*, pages 237–260. Chichester (UK): Wiley.

Rosenzweigh M. R. and Wolpin K. I. (2000). Natural "Natural Experiments" in Economics. *Journal of Economic Literature*, 38:827–874.

Rubin. (1980). Randomization Analysis of Experimental Data: The Fisher Randomization Test Comment. *Journal of the American Statistical Association*, 75(371):591–593.

Rubin D. B. (1986). Which Ifs Have Causal Answers (Comment on 'Statistics and Causal Inference' by Paul W. Holland). *Journal of the American Statistical Association*, 81:961–962.

Rubineau B., Lim Y., and Neblo M. (2019). Low Status Rejection: How Status Hierarchies Influence Negative Tie Formation. *Social Networks*, 56:33–34.

Russo F. and Williamson J. (2007). Interpreting Causality in the Health Sciences. *International Studies in the Philosophy of Science*, 21(2):157–170.

Sakoda J. M. (1971). The Checkerboard Model of Social Interaction. *Journal of Mathematical Sociology*, 1(1):119–132.

Salmon W. C. (1984). *Scientific Explanation and the Causal Structure of the World*. Princeton: Princeton University Press.

Saltelli A. (2000). What Is Sensitivity Analysis? In Saltelli A., Chan K. and Scott E. (eds.), *Sensitivity Analysis*, pages 3–13. Chichester (UK): Wiley.

Saltelli A., Chan K., and Scott E. (eds) (2000). *Sensitivity Analysis*. Chichester (UK): Wiley.

Sampson R. J. (2011). Neighborhood Effects, Causal Mechanisms and the Social Structure of the City. In Demeulenaere P. (ed.), *Analytical Sociology and Social Mechanisms*, pages 227–249. Cambridge: Cambridge University Press.

Sampson R. J., Winship C., and Knight C. (2013). Translating Causal Claims. Principles and Strategies for Policy-Relevant Criminology. *Criminology & Public Policy*, 12(4):587–616.

Sawyer R. K. (2003). Artificial Societies: Multiagent Systems and the Micro-Macro Link in Sociological Theory. *Sociological Methods and Research*, 31:325–363.

Sawyer R. K. (2004). Social Explanation and Computational Simulation. *Philosophical Explorations*, 7(3):219–231.

Schelling T. C. (1971). Dynamic Models of Segregation. *Journal of Mathematical Sociology*, 1:143–186.

Schupbach J. N. (2015). Robustness, Diversity of Evidence, and Probabilistic Independence. In Mäki U., Votsis I., Ruphy S. and Schurz G. (eds.), *Recent Developments in the Philosophy of Science: EPSA13 Helsinki*, pages 305–316. Heidelberg: Springer.

Shalizi C. R. (2006). Methods and Techniques in Complex Systems Science: An Overview. In Deisboeck T. S. and Kresh J. Y. (eds.), *Complex Systems Science in Biomedicine*, pages 33–114. New York: Springer.

Shoham Y. and Leyton-Brown K. (2009). *Multiagent Systems: Algorithmic, Game-Theoretic, and Logical Foundations*. Cambridge: Cambridge University Press.

Silverman E., Bijak J., Hilton J., Cao V. D., and Noble J. (2013). When Demography Met Social Simulation: A Tale of Two Modelling Approaches. *Journal of Artificial Societies and Social Simulation*, 16(4):9.

Simmons J. P., Nelson L. D., and Simonsohn U. (2011). False-Positive Psychology: Undisclosed Flexibility in Data Collection and Analysis Allows Presenting Anything as Significant. *Psychological Science*, 22(11):1359–1366.

Sims C. (1980). Macroeconomics and Reality. *Econometrica*, 48(1):1–48.

Skvoretz J. (2013). Diversity, Integration, and Social Ties: Attraction versus Repulsion as Drivers of Intra- and Intergroup Relations. *American Journal of Sociology*, 119:486–517.

Skvoretz J. (2016). All for One and One for All: Theoretical Models, Sociological Theory, and Mathematical Sociology. *The Journal of Mathematical Sociology*, 40(2):71–79.

Smith E. R. and Conrey F. R. (2007). Agent-Based Modeling: A New Approach for Theory Building in Social Psychology. *Personality and Social Psychology Review*, 11(1):87–104.

Smith J. A. and Burow J. (2020). Using Ego Network Data to Inform Agent-based Models of Diffusion. *Sociological Methods & Research*, 49(4):1018–1063.

Snijders T. A. B. (2014). Siena: Statistical Modeling of Longitudinal Network Data. In Alhajj R. and Rokne J. (eds.), *Encyclopedia of Social Network Analysis and Mining*. New York: Springer, https://doi.org/10.1007/978-1-4614-6170-8_312.

Snijders T. A. B. and Steglich C. E. (2015). Representing Micro-Macro Linkages by Actor-Based Dynamic Network Models. *Sociological Methods and Research*, 44:222–271.

Sobel M. A. (2006). What Do Randomized Studies of Housing Mobility Demonstrate?: Causal Inference in the Face of Interference. *Journal of the American Statistical Association*, 101(476):1398–1407.

Sobkowicz P. (2009). Modelling Opinion Formation with Physics Tools: Call for Closer Link with Reality. *Journal of Artificial Societies and Social Simulation*, 12(1):11.

Sorensen A. (1977). The Structure of Inequality and the Process of Attainment. *American Sociological Review*, 42(6):965–978.

Sørensen A. B. (1976). Models and Strategies in Research on Attainment and Opportunity. *Social Science Information*, 15(1):71–91.

Spirtes P. (2010). Introduction to Causal Inference. *Journal of Machine Learning Research*, 11:1643–1662.

Spirtes P., Glymour C., and Scheines R. (2000). *Causation, Prediction, and Search*. Cambridge, MA: MIT Press, 2nd edition.

Spirtes P., Glymour C. A., and Scheines R. (1997). Reply to Humphreys and Freedman's Review of Causation, Prediction, and Search. *British Journal for the Philosophy of Science*, 48(4):555–568.

Squazzoni F. (2012). *Agent-Based Computational Sociology*. Chichester (UK): Wiley.

Stadtfeld C. (2018). The Micro-Macro Link in Social Networks. In Scott R. A. and Buchmann M. (eds.), *Emerging Trends in the Social and Behavioral Sciences*. Wiley.

Steel D. (2004). Social Mechanisms and Causal Inference. *Philosophy of the Social Sciences*, 34(1):55–78.

Steel D. (2007). *Across the Boundaries: Extrapolation in Biology and Social Science*. Oxford: Oxford University Press.

Steglich C. E. G. and Snijders T. A. B. (2022). Stochastic Network Modelling as Generative Social Science. (Ch. 17) In Gërxhani K., De Graaf N. D. and Raub W. (eds.), *Handbook of Sociological Science. Contributions to Rigorous Sociology*. Cheltenham (UK): Edward Elgar.

Steglich C. E. G., Snijders T. A. B., and Pearson M. (2010). Dynamic Networks and Behavior: Separating Selection from Influence. *Sociological Methodology*, 40:329–393.

Steiger D. and Stock J. H. (1997). Instrumental Variables Regression with Weak Instruments. *Econometrica*, 65(3):557–586.

Stienstra K., Maas I., Knigge A., and Schulz W. (2020). Resource Compensation or Multiplication? The Interplay between Cognitive Ability and Social Origin in Explaining Educational Attainment. *European Sociological Review*, https://doi.org/10.1093/esr/jcaa054.

Stock J. H. (2001). Instrumental Variables in Statistics and Econometrics. In Smelser N. and Jab P. (eds.), *International Encyclopedia of the Social and Behavioral Sciences*, pages 7577–7582. Kidlington (UK): Elsevier Science.

Stock J. H. and Watson M. W. (2010). *Introduction to Econometrics*. Boston, MA: Addison Wesley, 3rd edition.

Stock J. H. and Yogo M. (2005). Testing for Weak Instruments in Linear IV Regression. In Stock J. H. (ed.), *Identification and Inference for Econometric Models*, pages 80–108. New York: Cambridge University Press.

Stock J. H., Wright J. H., and Yogo M. (2002). A Survey of Weak Instruments and Weak Identification in Generalized Method of Moments. *Journal of Business and Economic Statistics*, 20(4):518–529.

Stonedahl F. and Wilensky U. (2010a). Finding Forms of Flocking: Evolutionary Search in ABM Parameter-Spaces. Proceedings of the MABS workshop at the Ninth International Conference on Autonomous Agents and Multi-Agent Systems. Toronto, Canada.

Stonedahl F. and Wilensky U. (2010b). Evolutionary Robustness Checking in the Artificial Anasazi Model. In *Proceedings of the 2010 AAAI Fall Symposium on Complex Adaptive Systems*. Arlington, VA.

Sugden R. (2009). Credible Worlds, Capacities and Mechanisms. *Erkenntnis*, 70(1):3–27.

Sugden R. (2013). How Fictional Accounts Can Explain. *Journal of Economic Methodology*, 20(3):237–243.

Sugden R. (2000). Credible Worlds: The Status of Theoretical Models in Economics. *Journal of Economic Methodology*, 7:1–31.

Tesfatsion L. (2002). Agent-based Computational Economics: Growing Economies from the Bottom Up. *Artificial Life*, 8(1):55–82.

Tesfatsion L. (2006). Agent-Based Computational Economics: A Constructive Approach to Economic Theory. In Tesfatsion L. and Judd K. L. (eds.), *Handbook of Computational Economics. Agent-Based Computational Economics*, Vol. 2, pages 831–880. Amsterdam: North-Holland Elsevier.

Thiele J. C., Kurth W., and Grimm V. (2014). Facilitating Parameter Estimation and Sensitivity Analysis of Agent-Based Models: A Cookbook Using NetLogo and R. *Journal of Artificial Societies and Social Simulation*, 17(3):11.

Thorne B. C., Bailey A. M., and Peirce S. M. (2007). Combining Experiments with Multi-Cell Agent-Based Modeling to Study Biological Tissue Patterning. *Briefings in Bioinformatics*, 8(4):245–257.

Thorngate W. and Edmonds B. (2013). Measuring Simulation-Observation Fit: An Introduction to Ordinal Pattern Analysis. *Journal of Artificial Societies and Social Simulation*, 16(2):14.

Todd P. M., Billari F. C., and Simao J. (2005). Aggregate Age-at-marriage Patterns from Individual Mate-search Heuristics. *Demography*, 42:5559–5574.

Törnberg A. (2019). Abstractions on Steroids: A Critical Realist Approach to Computer Simulations. *Journal for the Theory of Social Behaviour*, 49(1): 127–143.

Van De Rijt A., Siegel D., and Macy M. (2009). Neighborhood Chance and Neighborhood Change: A Comment on Bruch and Mare. *American Journal of Sociology*, 114:1166–1180.

VanderWeele T. J. (2011). Sensitivity Analysis for Contagion Effects in Social Networks. *Sociological Methods and Research*, 40(2):240–255.

VanderWeele T. J. and An W. (2013). Social Networks and Causal Inference. In Morgan S. L. (ed.), *Handbook of Causal Analysis for Social Research*, pages 353– 374. Dordrecht: Springer.

VanHeuvelen T. (2018). Moral Economies or Hidden Talents? A Longitudinal Analysis of Union Decline and Wage Inequality, 1973–2015. *Social Forces*, 97(2):495–530. 10.1093/sf/soy045.

Varenne F. (2009). Models and Simulations in the Historical Emergence of the Science of Complexity. In Aziz-Alaoui M. A. and Bertelle C. (eds.), *i*, pages 3–21. Springer.

Vu T. M., Probst C., Nielsen A., Bai H., Meier P. S., Buckley C., Strong M., Brennan A., and Purshouse R. C. (2020). A Software Architecture for Mechanism-Based Social Systems Modelling in Agent-Based Simulation Model. *Journal of Artificial Societies and Social Simulation*, 23(3):1. http://jasss.soc.surrey.ac.uk/23/3/1.html.

Wang J., Zhang L., Jing C., Ye G., Wu H., Miao H., Wu Y., and Zhou X. (2013). Multi-scale Agent-based Modeling on Melanoma and Its Related Angiogenesis Analysis. *Theoretical Biology & Medical Modelling*, 10:41.

Wang X. and Sobel M. E. (2013). New Perspectives on Causal Mediation Analysis. In Morgan S. (ed.), *Handbook of Causal Analysis for Social Research*, pages 215–242. Dordrecht: Springer.

Watts D. J. (2014). Common Sense and Sociological Explanations. *American Journal of Sociology*, 120(2):313–351.

Weimer C., Miller J. O., Hill R., and Hodson D. (2019). Agent Scheduling in Opinion Dynamics: A Taxonomy and Comparison Using Generalized Models. *Journal of Artificial Societies and Social Simulation*, 22(4):5. http://jasss.soc.surrey.ac.uk/22/4/5.html.

Weisberg M. (2006). Robustness Analysis. *Philosophy of Science*, 73:730–742.

Wicherts J. M., Veldkamp C. L. S., Augusteijn H. E. M., Marjan B., Van Aert R. C. M., and Van Assen M. A. L. M. (2016). Degrees of Freedom in Planning, Running, Analyzing, and Reporting Psychological Studies: A Checklist to Avoid p-Hacking. *Frontiers in Psychology*, 7:1832.

Willer D. and Walker H. A. (2007). *Building Experiments: Testing Social Theory*. Stanford: Stanford University Press.

Williamson J. (2006). Causal Pluralism versus Epistemic Causality. *Philosophica*, 77(1):69–96.

Williamson J. (2013). How Can Causal Explanations Explain? *Erkenntnis*, 78(2):257–275.

Williamson J. (2015). Deliberation, Judgment and the Nature of Evidence. *Economics and Philosophy*, 31:27–65.

Williamson J. (2019). Establishing Causal Claims in Medicine. *International Studies in the Philosophy of Science*, 32(1):33–61.

Winsberg E. (2003). Simulated Experiments: Methodology for a Virtual World. *Philosophy of Science*, 70:105–125.

Winship C. (2009). Time and Scheduling. In Hedström P. and Bearman P. (eds.), *The Oxford Handbook of Analytic Sociology*, pages 498–520. Oxford: Oxford University Press.

Winship C. and Sobel M. (2004). Causal Inference in Sociological Studies. In Hardy M. and Bryman A. (eds.), *A Handbook of Data Analysis*, pages 480–504. London: Sage Publications.

Woodward J. (2002). What Is A Mechanism? A Counterfactual Account. *Philosophy of Science*, 69:S366–S377.

Woodward J. (2003). *Making Things Happen. A Theory of Causal Explanation*. New York: Oxford University Press.

Wooldridge M. (2000). *Reasoning about Rational Agents*. Cambridge, MA: MIT Press.

Wooldridge M. (2009). *An Introduction to Multi Agent Systems*. Chichester (UK): Wiley.

Wunder M., Suri S., and Watts D. J. (2013). Empirical Agent Based Models of Cooperation in Public Goods Games. In *Proceedings of the Fourteenth ACM Conference on Electronic Commerce*, EC '13, pages 891–908. New York: ACM.

Wurzer G., Kowarik K., and Reschreiter H. (eds.) (2015). *Agent-Based Modeling and Simulation in Archaeology*. Berlin: Springer.

Xia H., Wang H., and Zhaoguo X. (2011). Opinion Dynamics: A Multidisciplinary Review and Perspective on Future Research. *International Journal of Knowledge and Systems Science*, 2(4):72–91.

Ylikoski P. (2013). Causal and Constitutive Explanation Compared. *Erkenntnis (1975-)*, 78:277–297.

Ylikoski P. and Aydinonat N. E. (2014). Understanding with Theoretical Models. *Journal of Economic Methodology*, 21(1):19–36.

Young P. (2006). Social Dynamics: Theory and Applications. In Judd K. and Tesfatsion L. (eds.), *Handbook of Computational Economics*, Vol. II, chapter 22. Amsterdam: North-Holland.

Zachrison S. K., Iwashyna T. J., Gebremariam A. et al. (2016). Can Longitudinal Generalized Estimating Equation Models Distinguish Network Influence and Homophily? An Agent-based Modeling Approach to Measurement Characteristics. *BMC Medical Research Methodology*, 16:174. https://doi.org/10.1186/s12874-016-0274-4.

Zhang L., Chen L., and Deisboeck T. (2009). Multiscale, Multi-resolution Brain Cancer Modeling. *Mathematics and Computers in Simulation*, 79(7):2021–2035.

Ziliak S. T. and McCloskey D. N. (2008). *The Cult of Statistical Significance: How the Standard Error Cost Us Jobs, Justice, and Lives*. Ann Arbor, MI: University of Michigan Press.

Index

Agent-based Models and Causal Inference, First Edition. Gianluca Manzo.
© 2022 John Wiley & Sons, Inc. Published 2022 by John Wiley & Sons, Inc.